Reproductive Genetics, Gender and the Body

This book is about 'reproductive genetics', a sociological concept developed to define the use of DNA-based technologies in the medical management and supervision of reproduction and pregnant women. In a searching analysis, it uncovers the powerful but hidden social processes involved in the development of these technologies.

Focusing on prenatal screening, Elizabeth Ettorre explores how the concepts of gender and the body are intertwined with the process of building genetic knowledge, and some of the unintended consequences for women. These include the development of a gendered discourse of shame and stigmatisation in which the 'perfect' body becomes idealised and new conceptions of 'disability' are formed. Drawing on interviews with European medical, legal and nursing professionals, she shows how the modernist tradition of scientific disinterestedness is being replaced by a new ethic: the making of moral judgements by scientists.

Reproductive Genetics, Gender and the Body raises important issues around the gendered, female body – the site of genetic capital. It challenges professional and academic alike to construct new perceptions about pregnant women and their embodied experiences of reproductive medicine.

Elizabeth Ettorre is Professor of Sociology at the University of Plymouth.

Reproductive Genetics, Gender and the Body

Elizabeth Ettorre

London and New York

First published 2002
by Routledge
11 New Fetter Lane, London EC4P 4EE

Simultaneously published in the USA and Canada
by Routledge
29 West 35th Street, New York, NY 10001

Routledge is an imprint of the Taylor & Francis Group

Typeset in Times by Taylor & Francis Books Ltd
Printed and bound in Great Britain by TJ International Ltd, Padstow,
Cornwall

British Library Cataloguing in Publication Data
A catalogue record for this book is available from the British Library

Library of Congress Cataloging in Publication Data
A catalogue record for this book has been requested

ISBN 0–415–21384–3 (hbk)
ISBN 0–415–21385–1 (pbk)

To my dear Finnish sister, Arja Laitinen, who made me remember that when the time comes, biomedical knowledge can't save any of us from death.

Contents

Acknowledgements

I wish to thank the publishers for granting permission to reprint parts of the following:

(2000) 'Reproductive genetics, gender and the body: "Please doctor, may I have a normal baby?"', *Sociology* 34, 3: 403–20.

(1999) 'Experts as genetic storytellers: exploring key issues', *Sociology of Health and Illness* 21, 5: 539–59.

(1998) Review article: 'Reshaping the space between bodies and culture: embodying the biomedicalised body', *Sociology of Health and Illness* 20, 4: 548–55.

(1997) 'The complexities of genetic technologies: unintended consequences and responsible ethics', *Sosiaalilääketieteelinen Aikakauslehti* (Journal of Finnish Social Medicine) 34: 257–267.

I am grateful to my European research partners for arranging my interviews with experts in their countries. I am also grateful to my expert respondents who gave me their time out of very busy schedules.

My appreciation goes to my colleagues at the Department of Sociology, University of Plymouth for granting me sabbatical leave (from January 2001–September 2001) so that I was able to complete this book.

Thanks goes to Heather Gibson, formerly at Routledge, who commissioned this work. Also, I would like to express a special thanks to Edwina Welham, my editor at Routledge, who has been particularly patient. (I moved house, jobs and country during the writing of this book so there have been some delays.)

And to Irmeli of course, many many thanks.

Introduction

The sociology of reproductive genetics: the institutions of reproduction and gender and genes in bodies

'Humanity is unnatural!' exclaimed the philosopher Dynyasha Bernadettesion (A.C. 344–426) who suffered all her life from the slip of a genetic surgeon's hand which had given her one mother's jaw and the other mother's teeth...

Joanna Russ, *The Female Man*, p. 12

Introduction

This book focuses on reproductive genetics, which I define as the utilisation of DNA-based technologies in the medical management and supervision of reproduction, and ultimately, female bodies. Although this concept, reproductive genetics, may appear at first glance as biomedical, it is not. Reproductive genetics is a sociological concept employed to demonstrate that powerful social and cultural processes are involved in the medical organisation of genetic tests for prenatal diagnosis, already identified as an intricate sociotechnological system (Cowan 1994: 35).

In feminist contexts, this concept, reproductive genetics, has been used to demonstrate that it is not possible to treat women and men in the same way with regard to reproduction (Mahowald 1994: 69). From the standpoint of women and people with disabilities, it is on this wider social plane that the moral stakes are highest (Faden 1994: 88). In this context, Wendell (1992: 71) argues that reproductive genetics represents one of the most successful achievements of modern medicine to control 'nature'. For her, reproductive genetics is a powerful strategy within the Western scientific project to idealise the 'perfect' body and thus shape contemporary conceptions of able-bodiedness and disability.

The development of reproductive genetics is the process of formation of networks of relatively stable ensembles of procedures, instruments, theories, results and products to which various actors give their allegiance. With this development, prenatal technologies are used increasingly for foetal diagnosis. The procedures, which can be either non-DNA-based or DNA-based, will be discussed fully in the following chapter. Prenatal screening and diagnosis are complicated medical processes which involve numerous tests, myriad social relations with a variety of medical personnel (for example, obstetricians, gynaecologists, genetic counsellors, midwives, paediatricians, etc.) and complex decision-making processes, extending for some women, over several pregnancies (Parsons and Atkinson 1993). Prenatal screening and diagnosis of foetuses for congenital conditions allows physicians to perform selective abortion of 'affected' foetuses. For them, a specific genetic technique may appear as one of many solutions to the problems of disability and disease that modern reproductive medicine seeks to solve. Any perceived costs such as spontaneous or selective abortion, negative psychological effects, and physical pain are generally viewed as negligible in relation to the perceived technological benefit such as the birth of 'normal', non-disabled babies. Professionals may see these technologies as helping to detect foetal abnormalities, while pregnant women tend to see them as an assurance that no abnormality has been detected (Green 1990a).

In this context, Farrant (1985) argues that these sorts of technologies have not been developed with the interests of pregnant women primarily in mind. Yet, the perceived biomedical wisdom that exists is that pregnant women are helped by prenatal genetic testing. Along with other feminists (Katz Rothman 1998a, 1998b, 1996, 1994; Markens *et al.* 1999; Hallebone 1992; Corea 1985; Corea *et al.* 1985), I want to challenge this view and show how it is merely one side of the equation.

The institutions of gender and reproduction

Traditionally, gender has been viewed as society's expectation concerning behaviour viewed as appropriate for members of each sex, male and female. Sociologists have explored gender as both a complex social process and an established institution (Lorber 1994). For example, as a process, gender is embedded in all human interactions. Gender shapes the meaning of female and male as well as representations and performances of masculinity and femininity on cultural, political and economical levels. Gender has an effect on the social groupings of men and women and divisions between both the private and public arenas of social life.

As an institution (Lorber 1994), gender is a part of culture just like other components of culture such as symbols, language, values and so on. Gender is a complex form of structured inequality, embedded in our daily lives. Gender is a moralising system. It embeds into society a set of inter-related norms centred on the activities of individuals and prescribes these activities. These individuals are marked by differences on the basis of being male and female as well as masculine and feminine. In this book, I use the word, gendering to emphasise the varied, complex social processes that embed gender in our diverse cultures. Gendering processes are institutionalising processes.

Here, I would contend that similar to gender, reproduction as a component of culture is exhibiting signs of a social institution. As reproduction ascends as a social institution, it develops into a regulatory system, focused on the replication of bodies which must exemplify completeness (i.e. organs, limbs, torsos, crania filled with brains, etc.), health, well-being, individual potential and future welfare. Reproduction represents an organisation of values, norms, activities and social relations that symbolise notions of able-bodiness, human survival, progress and individual potential. At the same time, reproductive bodies, especially female reproductive bodies, become more valorised than ever before through the surveillance of their pregnant wombs by medical specialists' use of prenatal technologies. These technologies can be seen to be in 'diffusion', meaning that they are being routinely employed within health care systems and firmly entrenched in the social matrix (Cowan 1994). From this viewpoint, reproduction is a normative, standardising system, which disciplines, controls and scrutinises the actions of procreative bodies, both male and female. But, female more than male bodies are constructed as reproductive and female procreative bodies tend to be subject to more medical surveillance, control and 'technologising' than the procreative bodies of males. Given that the long process of conception and gestation is internal to the female body, the reproductive body stands for something essentially female (Shildrick 1997: 22).

There is a multiplicity of practices, guiding pregnant bodies as these gendered bodies are gathered together in a systematic way under the flag of reproduction. The symbol of reproduction as an emergent social institution is the pregnant body, the body of a woman producing a baby, as well as the chemicals, hormones, eggs, cells, genes, blood, foetal tissues, all gathered, drawn, scraped, tested, examined, and at times, discarded within reproductive medicine. An assortment of disciplinary strategies (i.e. biomedical knowledge, technologies, etc.) attends to the pregnant body to construct and normalise it. This process is carried out under the supposedly, benevolent gaze of the physician.

The massive proliferation of genetic technologies into reproductive medicine calls for an examination of the complexities of these technologies, how they have developed thus far and the consequences, both intended and unintended, of these technologies on modern social life. For Western culture, the practice of medicine and the use of medical technologies have developed within medical systems that are articulated as cultural systems (Kleinman 1991). It is within these medical systems with their own meanings, values and behavioural norms that reproductive genetic technologies have emerged. These technologies have grown from within a medical system and thus, scientific culture whose members have identified traditionally with social progress and humane goals (Foucault 1973; Turner 1987). Nevertheless, genetic technologies in the field of reproduction are not in themselves benevolent techniques.

In *Reproductive Genetics, Gender and the Body*, I want to explore the complexities of genetic technologies with special reference to biomedical prenatal practices. I also want to establish gender and the body as key contextual and theoretical sites in this many layered exploration. The main aim of this book is to show how the field of reproductive genetics can benefit from a feminist or critical gender perspective and how gender, the body and ethics go hand in hand. My assumption here is twofold. Firstly, genetic technologies, underpinning prenatal technologies, are most significantly socially and culturally shaped. Prenatal genetic technologies are not merely technical, biomedical artefacts. They include the set of relations between the artefacts and their surrounding actors, assumptions and practices (Rudinow Saetnan 1996). An analysis of these technologies includes an examination of the specific techniques and procedures used in the prenatal area, the whole process of building genetic knowledge and the mode of work that accompanies the development of these specific genetic techniques. These dimensions of prenatal genetic technologies – technique, knowledge and organisation – need to be assessed in order to understand them fully.

Secondly, these technologies have unintended consequences which remain invisible in the field of reproductive genetics. While there are disparate discourses concerning the desirability of these technologies amongst feminists and pregnant women themselves (Lupton 1994), these technologies have been found to have less than calming effects upon pregnant women (Green 1990b). I would contend that there is a need for analyses of gender relations and corporeal dynamics in this field. To understand the importance of developing a feminist perspective is to recognise the powerful interplay between the field of genetics, modern reproductive medicine, foetal diagnosis and pregnant bodies. In this way, I hope that this book will not only contribute new insights into a field that has been popularised by the professional and public media (Laurén *et al.* 2001) but also provide a fresh understanding of increasingly familiar material by treating it in an innovative way.

Technology, gender and broken bodies

Over the past three hundred years, biomedical scientists have owned, developed and managed the study of the body. They have been the main proselytisers on how this 'machine' works. Indeed, the biomedical discourses on the body have become entrenched in contemporary cultures as our bodies have been, time and again, shaped by notions embedded in Cartesian dualism. Living bodies have been treated as no different from a piece of equipment, while this powerful and far-reaching discourse has consistently obscured considerations of sentient bodies. In recent years, sociologists (Turner 1996, 1992; Frank 1995; Williams and Bendelow 1998; Nettleton and Watson 1998) have begun to position bodies centrally to studies in sociology, specifically the sociology of health and illness. At the same time, specialists in biomedicine continue to invent ways to make bodies more accessible for their perusal, while altering the boundaries of these bodies through various procedures such as organ transplants, limb and skin grafts and through exploring more possibilities of xenotransplantation.

Bryan Turner (1992: 165–6) contends that the uncertain status of the body in human cultures, the contradictory relationship between nature, society and the body and the social role of illness in human cultures acts as a symbolic map of the political and social structure are all the province of medical sociology. Indeed, bodies are growing to be more compliant and more disciplined as they progress from being healthier or sicker.

While the concepts of health and illness are culturally and socially defined, all cultures have known concepts of these terms. These terms differ from culture to culture in relation to how ailing and well bodies become visible as well as the intensity of the 'scopic drive'. The scopic drive is the biomedical quest to make the unseen visible in the biotechnological world (Braidotti 1994). Whether sick or healthy, bodies are viewed as empirical objects to be quantified, classified, visualised and disciplined through biomedicine. Thus, through this biomedical gaze, bodies are treated as 'things' to be studied and not as embodied subjects.

Alongside this scopic drive produced by biomedicine, an authoritative need to categorise embodiments of the 'abnormal', 'irregular', 'odd' or 'deviant', situate them in biology and patrol these bodies in public spaces exists in society. Indeed, Urla and Terry (1995: 1) argue that since the nineteenth century, the somatic territorialising of deviance has been an important component of a larger effort of the State including scientists, lawmakers and the police to organise social relations according to categories denoting health versus pathology, normality versus abnormality and national security versus social danger. As a result of these multifarious practices, bodies have become marked accordingly and social relations organised in terms of deviant and normal bodies. Nevertheless, before cultural conceptions of normal and abnormal, conformity and non-conformity, or health and pathology can be constructed, an assemblage of bodies upon which these conceptions can be inscribed needs to exist. Specifically, qualified biomedical experts using their convincing discourses and scientific practices perform the distinctive process of inscribing bodies as healthy or diseased. This process of inscription includes the pressing of technology into the service of medicine.

Modern biotechnology has been described as an ambiguous mix of knowledge and engineering, science and technology, nature and culture, possibilities and risks, hopes and fears (Nielsen 1997: 102). Regardless of the fact that technological medicine produces simultaneously great expectations as well as more arenas for uncertainty (Freund and McGuire 1999: 222), the union between the medical profession and biotechnology has empowered the profession as well as helped to create lucrative returns on many biotechnological investments, especially in the global genomics industry (Rabinow 1999). As the concept of biotechnology developed, it was seen to integrate the contemporary idea of manufacturing with visions of humanity (Bud 1993: 52). The stuff of biology became raw material for medical exploration and in some instances, exploitation and death (Jones 2000). With regards to the replication and production of bodies, biotechnology has had a major impact on present-day notions of reproduction. Wajcman (1991) argues that in no other area of social life is the relationship between gender and technology more forcefully disputed than in the sphere of human biological reproduction. Indeed, 'technological fixes' have been increasingly applied to pregnant, female bodies and some of these pregnant bodies, I would argue, emerge as 'broken bodies'.

In using this term, 'broken bodies', I want to flag up the idea that our Western ways of thinking about the body have been based on separating ourselves as social and moral actors from our bodies.

Simply, we have become disembodied. Furthermore, because morality is highly mediated by gender, morality has been based on the exclusion of female bodies from full moral agency. Through moral agency one experiences oneself as a contributor in the process of evolution toward greater autonomy and connectedness (Tomm 1992: 108). Clearly, women have lost out. Furthermore, any person whose unique experiences have been largely omitted from the dominant culture (e.g. poor, ethnic minorities, lesbians and gay men, the disabled, etc.) will feel this loss. It is a distortion to think that generic persons exist in moral situations or that gender is irrelevant to moral deliberations (Warren 1992). Making morals is an embodied experience as well as a deeply gendered process.

Women experience a fragmented morality of the body. Simply, in moral terms, their bodies are not whole; they have become broken. Sharp (2000) argues that the application of reproductive technologies and other biotechnologies mark a paradigmatic shift in our understanding of the body. Bodies become commodified and fragmented. Furthermore, a hidden morality surrounds pregnant bodies. On the one hand, these pregnant bodies are appropriated into a series of gendered narratives – mythological, biblical, classical humanist and anthropological. On the other hand, these gendered narratives help to discipline the cultural threat posed by reproductive bodies (Newman 1996) and are hostile to female subjectivity (Klassen 2001). We need to challenge this type of hidden morality and ensure that pregnant bodies remain whole, sentient bodies.

Bringing whole bodies back in to sociology has been made possible by the efforts of women bringing themselves back in. Academic feminists have exposed that the traditional neglect of the body reflected a masculinist sociology that naturalised bodies and furthermore, legitimated control of male over female bodies. Feminists have documented the types of regulation, restraint, provocation and resistance experienced by gendered bodies. Anne Witz (2000: 2) contends that our disembodied sociological heritage includes a history of 'her excluded body' and 'his abject body'. She cautions that the recuperated body in sociology is in danger of being the abject male body; a warning that is heeded in this book.

While recognising the need to recover the lived experiences of both the excluded and abject body in sociology, I recognise that all sorts of activities in which we are involved as social beings are embodied activities. Similar to Arthur Frank (1995), I would suggest that sociologists become aware of the fact that that what bodies experience, suffer, bear, desire and consume should be the foundation stones for sociology.

Here, I would add that these embodied experiences should also enlighten a sociology of reproductive genetics.

In this book, I want to make sure that the stories I tell reflect the lives of the people I study. I see this type of professional behaviour as ethical work. Arthur Frank (1991, 1995) has argued that there is a need for an ethics of the body, shaping a sociology of the body. He equates ethics with a social science that empathises with people's suffering. For him, only an empathic social science that witnesses suffering is worthy of our attention. He wants sociologists to give careful consideration to bodies in order to bear witness to what people suffer. So for me as a feminist sociologist interested in reproductive genetics, ethics means offering true reflections that are empathetic as well as attentive to reproductive bodies. Ethics means that I do not consider these issues in a gender, race and class neutral manner. Rather, I argue that considerations of these differences and other differences (sexuality, ability, age, etc.) among individuals in reproductive genetics are crucial to maintain an ethics of the body as well as uphold care and justice (Mahowald 1994: 67) in society. As embodied beings, we put the human body on a pedestal. When will we develop an ethics of the body that is attentive to our diverse, embodied realities?

On another level, ethics is defined as 'rules of conduct'. But ethics can also be defined as 'the science of morals' (*Concise Oxford Dictionary* 1995). From a feminist point of view, care must be taken when we speak of ethics in the context of women's embodied moral lives, given that contemporary social practices encourage discrimination against us and the suppression of women's moral views (Browning-Cole and Coultrap-McQuin 1992). Here, a central strand of thinking, informing the development of the ideas presented in this book is that all advancements within reproductive genetics should be framed by feminist perspectives on bioethics (Holmes and Purdy 1992).

Viewed as ethical, beneficent medical practices, regulatory techniques are played out on pregnant bodies through the science of reproductive medicine. Within biomedical knowledge, benevolence is the virtue of being disposed to act for the benefit of others, while beneficence is the moral obligation to do so. Both terms can be linked to paternalism (Beauchamp and Childress 1994). But, when the pregnant body enters into this moral equation and becomes a theoretical site, problems can be seen to exist. For example, physicians may override pregnant women's autonomy *vis-à-vis* prenatal genetic technologies and justify their actions by the goal of avoiding harm – the birth of a disabled baby. On another level, reproduction increasingly comes to be constructed as a matter of consumption and the foetus a commodity (Taylor 2000). Here, medical paternalism, bureaucracy and capitalism make an easy mix.

Medicine needs healing and the feminist project of healing medicine utilises 'epistemic empathy', offering oppressed groups help and insights based on gender sensitive theories and practices (Holmes 1992: 3). Certainly, the practice of mixing prenatal techniques with the judgements of individuals, including both physicians and their female patients, demands a rigorous, moral formula upon which future research and service development can be based (SGOMSEC 1996).

Spallone *et al.* (2000) argue that the use of prenatal genetic techniques demands a socially informed ethics which provides a way of allowing a sense of social responsibility, rooted in an understanding of the effects of new technologies, to replace the one-dimensional requirement for quality control and technical expertise. Given the above, I would contend that an ethics of the body and feminist ethics are interrelated.

Seeing genes in bodies as a political process

When individuals are reduced to their genetic codes, experts define disorders, behaviours and physiological variations as in part genetic in origin. They may advocate genetic technologies as interventions to manage a variety of social issues, masked as 'health' problems. Abby Lippman (1992) is concerned about the increased use of genetics to solve social problems, which she calls 'geneticisation'. Lippman argues that through this process of geneticisation, individuals are condensed to their DNA codes. She wants social scientists to reclaim the genetics agenda and expose the colonisation of health and illness through biomedical narratives of genetics.

Carols Novas and Nicholas Rose (2000) contend that Lippman's geneticisation argument implies that ascribing genetic identity to individuals and groups objectifies them, denying them their human agency. In their view, to make humans the object of genetic knowledge is not subjection but the creation of subjects. These authors contend that genetic knowledge can empower some people who are at genetic risk and establish them in a network of genetic connectedness. While they acknowledge that prenatal genetic testing may open up new strategies of control, the authors believe that the reconfiguration of identity within genetics establishes a new ethical field for the formulation of life strategies. They argue that life strategies are put together in a complex field of ongoing ethical problematisation of how one should conduct one's life. But, questions arise if these ideas are applied to reproductive genetics: Who will define this new ethical field: the pregnant woman as active subject/patient or the professional? If new strategies of control are practised upon pregnant bodies, do these strategies have an impact on how the active subject/patient (e.g. pregnant woman) reconfigures her identity or is disciplinary power suspended in this instance?

I agree with Abby Lippman. Social scientists should reclaim the genetics agenda and demonstrate how human bodies – genes and all – are politically shaped. Our bodies, biologies and pedigrees are shaped by social practices of repression and control as well as social exclusion. Class, race, ethnicity, gender, age, disability and nationality shape public perceptions about human bodies and these powerful perceptions affect people's lives. Think of disability. If someone is disabled, she or he is not only viewed as either physically or mentally impaired, but also suffers from society's dislike of undesirable physical, sensory or mentally related difference.

The culture of medicine and specifically, the professional practices of physicians with regards human reproduction have changed dramatically with the rapid technological developments, generated by geneticists and molecular biologists. The contribution of genetic factors to the entire range of diseases is being recognised, while the public hears that a greater proportion of childhood disease is genetic in origin (Clarke 1997a: 5). However, the medical profession's use of genealogical pedigrees to demonstrate that heredity was involved in the aetiology of a particular illness or pathological condition can be traced to the mid-nineteenth century (Resta 1999). In many countries, medical experts contend that all pregnant women should undergo prenatal diagnosis and be screened for genetic disorders. At the same time, men and women in the general population believe that these sorts of techniques can be helpful in eliminating many serious diseases. In my work on reproductive genetics, I see pregnant bodies as 'gendered' sites where knowledge of genes, foetuses, reproductive processes and biomedicine converge. Pregnant bodies are not gender-neutral systems. One difficulty with this work is to show how ideas about 'genetic codes in bodies'; models about healthy and diseased genes; and data about appropriate kinds and levels of reproductive performances are culturally dependent, 'embodied processes'. I am concerned in this book with affirming bodies: making the distinct claim that pregnant bodies exist very centrally in the reproductive genetics discourse.

Here, to understand the importance of developing a feminist perspective is to recognise the powerful interplay between the field of genetics, modern medicine and the bodies of women. Is it fair that some women condemn themselves to shame because they do not measure up to society's image of what it means to be a good reproducing mother? Is it just that they experience their bodies as broken? The obvious answer is no. But, that these questions need to be asked demonstrates that prenatal genetic testing has profound ethical implications. If women's reproductive processes continue to be ranked according to genetic information, the field of genetics needs to become more gender sensitive than it is at present and we need to see that gender, bodies and ethics go hand in hand.

Here, it should be noted that this book will not focus on the new reproductive technologies (NRTs) which I term 'pre' prenatal technologies (i.e. in terms of the site for foetal development) which include a variety of techniques. For example, *in vitro* fertilisation (IVF) in conjunction with superovulation, ultrasound, laparoscopic egg retrieval and embryo transfer as well as gamete intrafallopian transfer (GIFT) have allowed infertile women or women beyond childbearing age to experience pregnancy, if not give birth to a child. There has been work, written specifically by feminists (see for example Edwards *et al.* 1999; Franklin 1997; Steinberg 1997), which addresses the problems related to 'conception assisting' NRTs for infertile women. While genetic technology is related to NRTs (i.e. the genetic composition of eggs, sperm and embryos are monitored before implantation), prenatal screening and diagnosis, on the other hand, tends to be focused on fertile, already pregnant women. All of these practices can be seen as assisting the birth of a 'normal', 'non-afflicted' baby/child.

Within the women's health movement, the NRTs have been criticised (Arditti *et al.* 1984; Corea 1985; Stanworth 1987; Spallone and Steinberg 1987a; McNeil 1993) along with the overall nature, context and management of reproduction within the medical profession (Graham and Oakley 1991). The NRTs have had manifold implications on our scientific and popular conceptions of motherhood (Lewin 1985; Stacey 1988). Some criticisms point to basic ethical issues (Overall 1987; Elshtain 1991) as well as to some of the burdens which most, if not all, NRTs place upon women's bodies and psychologies (Hallebone 1992; Corea *et al.* 1985; Katz Rothman 1994; Tymstra 1991; Marteau 1989).

The source of empirical data presented in this book is a qualitative study on experts' accounts of the use of prenatal genetic technologies in four European countries: England, Finland, the Netherlands and Greece from 1996–9. This study comes from a European consortium of researchers carrying out a series of seven comparative studies in this area (see Ettorre 2001a).

Methods for the experts' study

The sample

Given that the aim of the experts' study was to review the role of key players influential in public debates about reproductive genetics in each country, I wanted to find experts who were active in this area either clinically or academically. Initially, I hoped to interview at least ten experts in each country. The goal was to interview equal numbers of geneticists, clinicians, practitioners, lawyers and/or ethicists, policy makers, public health officials and researchers. While the study focus was on biomedical not lay experts, this is not meant to imply that only professionals can be experts, a view that can be disputed (Calnan 1987, Davison *et al.* 1994). It was also decided that besides being known as an 'expert', a prerequisite for inclusion in the study would be fluency in English.

A potential sample of respondents was selected from a list of ten known experts drawn up by researchers (the European research partners) in each of the four countries. Experts were known through their publications, work contacts and/or national reputation within these researchers' networks. A final list for inclusion in the study was discussed and drawn up jointly by the author (who would carry out the interviews) and country researchers.

Data collection and analysis

After experts were selected, they were contacted by country researchers and asked whether or not they would be interested in providing their views. With two exceptions, all experts who were selected agreed to be interviewed. With the help of the local researchers, the author set dates when she would be visiting each country as well as arranged times and places for the interviews. Twenty-eight interviews were completed between October 1996 and December 1996. Before data collection began, I decided that in order to keep to the study timetable, an additional fourteen interviews carried out during the Finnish pilot study in 1995 would be included in the main study. This meant that the final number of interviews was forty-five. These included seventeen both pilot and full-study interviews for Finland (one gynaecologist, seven medical geneticists, four policy makers, two public health officials, one researcher and two lawyers); ten for Greece (a paediatrician, an obstetrician, a clinical geneticist, two policy makers, three lawyers and two ethicists); nine for England (two medical geneticists, an epidemiologist, two policy makers, one public health official, two researchers and an ethicist) and nine for the Netherlands (a general practitioner, three obstetricians, a medical geneticist, a policy maker, a midwife, a researcher and an ethicist). The interviews were conducted in English, tape-recorded and lasted between thirty to ninety minutes: the average was an hour.

Here, it should be noted that although it was possible to reach the desired numbers of interviews in Finland and Greece, this was not possible in the Netherlands and England. Both countries were one interview short. This was due mainly to the study timetable and budget. All interviews needed to be completed by December 1996 and the researcher had one week's time to interview experts in each country. To account for this lack, seven 'unofficial' interviews (i.e. without the use of a tape recorder) were carried out: four in the Netherlands and three in England. These seven 'unofficial' interviews were mainly with medical students or biomedical researchers with expertise in the area. The data obtained from these 'unofficial' interviews were not used in the data analyses and were perceived by the author as providing background information in specific countries.

Experts were asked their views on prenatal genetic testing and screening. The interview questions revolved around their attitudes on the use of these techniques; their perceptions of the prevailing state of knowledge on legal, medical and ethical aspects; social effects; and policy priorities on local and national levels.

While I spent a week in each country, my visits tended to be 'flying visits' with me rushing around a particular country using taxis, trains and buses to reach my interview appointments on time. Sometimes, I would wait outside of buildings if I arrived early checking that my tape recorder was working and seeing colleagues and patients arrive in clinics.

During my visits to clinics, hospital departments, medical academic departments, government offices, general practices, medical organisations, and so on, I did not meet patients with one exception. In my mind, not meeting patients was appropriate. I came to these various places to interview the professional treater not his/her patients. The one exception was when I actually met a patient while I was interviewing a midwife at a general practice. Just after our interview started, the midwife had a phone call. She then interrupted our interview and said something in her language down the phone. Shortly after she put down the phone, the door opened and a very pregnant woman came in. At first, I was confused and looked at the midwife. She said something to the woman in her language. Then in English, she introduced me to the patient, Marta (a pseudonym), and said that

I was here to interview her as a midwife. The midwife then said that Marta was due to go on holiday that day and was worried because she had not felt the foetal heart beat. The midwife said that she would just check for Marta and it would take only a minute after which we would resume the interview. I said fine that I would leave. She said: 'No just sit down here because there is no where for you to go. All the rooms are engaged.' So I just waited uncomfortably because I felt that I should not be there. In the end, the consultation with Marta took about four minutes and she left a happy customer. I do not think it was appropriate that I was in that room, but there was no other room for me to go in that general practice. There was not even a waiting room. This did not happen again in this study because most of the experts that I saw after the midwife were not seeing their patients on the day of the interview.

All interview data were transcribed and key themes were identified by Word Perfect (Windows) Quick Finder Index. Subsequently, these themes were discussed amongst the research partners. Excerpts from the experts' interviews will be used in this book. I should mention here that some variations of experts' attitudes occurred within specific disciplines (e.g. clinical geneticists disagreed with other clinical geneticists), amongst disciplines (e.g. differences between clinical geneticists, obstetricians, ethicists, policy makers) and between countries (e.g. differences between Dutch and English experts). With regards the last issue, when cross cultural differences are emphasised in this area, we tend to see less clearly the sorts of inter-disciplinary disagreements (such as those between clinical geneticists and policy makers, etc.), which are taken for granted in single country studies (Kerr *et al.* 1997; Gilbert and Mulkay 1984). The focus in this book is mainly on similarities in experts as a group and their attitudes towards reproductive genetics. However, reporting on cross-cultural variations will be presented in Chapter 6.

The chapters in this book will contain excerpts from interviews conducted in this study. All interviews are numbered in chronological order. F stands for Finland, E for England, G for Greece and NL for the Netherlands. For example, E 1 indicates the first interview carried out in England.

The structure of this book

Chapters 1 through Chapter 3 explore the complexities of the social shaping of reproductive genetics by employing a multi-dimensional analysis of prenatal genetic technologies. This type of analysis has been based on the assumption that prenatal genetic technologies include specific techniques and procedures (Chapter 1), and the whole process of building genetic knowledge (Chapter 2) and the mode of work which accompanies the development of specific prenatal genetic techniques (Chapter 3). Thus, the technologies used in reproductive genetics are viewed as the culmination of experimental techniques; biomedical knowledge; the knowledge interests of scientists, geneticists, researchers, physicians and so on; and the observable method of work, accompanying these techniques within the current state of thinking.

You, the reader, are invited to look at the relevant, experimental techniques and procedures with a critical eye in Chapter 1, 'Prenatal Politics and "Normal Patient Families"'. This chapter focuses on the commodification of gendered bodies through the use of special techniques of reproductive genetics including the mixture of chemical, drugs, devices, medical and surgical procedures and gendered bodies. After a description in detail of these procedures/techniques and their development, the contention is made that regardless of the constant profusion of these techniques, genetic and indeed prenatal genetic technologies must be seen as more than techniques. These technologies are the extension of human competence, proficiency and skill into the science of genetics, a science located squarely within the workings of the medical system. In the final discussion of this chapter, the relationship between reproductive genetics and 'normal patient' families is explored.

In Chapter 2, 'Biomedical Knowledge and Interests: Genetic Storytellers and Normative Strategies', a critique of the knowledge base of genetic technologies with special reference to organisation of knowledge developed around the space defined as 'prenatal' is presented. The notion of experts as genetic storytellers will be presented along with the sorts of normative strategies they use in disseminating genetics knowledge. They are producers of genetic ideology and have a great impact on the field of reproductive genetics. Genetic storytellers possess particular knowledge interests concerning the role of genetics in society.

In Chapter 3, 'Organisation of "Genetics Work": Surveillance Medicine and Genetic Risk as a Novelty', two related questions will be raised: how is the work of predictive genetics in the prenatal area organised within biomedicine, the medical arm of genetics? and, what is happening within modern medicine to aid in the expansion of genetic technologies and in turn, prenatal technologies? In this chapter, genetic risk identity is viewed as novel and the contention is made that social scientists should reclaim the genetics agenda and challenge the current misconceptions shaped by the discipline of biomedicine.

Chapters 4 and 5 include discussions of some of the unintended consequences of reproductive genetics and its effects upon female bodies. My intention in these two chapters is to develop a feminist embodied approach to reproductive genetics and generate an awareness the disciplinary practices in which pregnant bodies, seeking healthy, if not 'perfect', babies are involved. What happens when the pregnant body becomes a theoretical site?

Chapter 4, ' Shaping Pregnant Bodies: Distorting Metaphors, Reproductive Asceticism and Genetic Capital', emphasises that the discourses supporting reproductive genetics become very powerful ways of shaping the experiences of pregnant bodies. Genetic metaphors help to create distorted ideas of the human body as well as render invisible bodies inscribed by categories of difference. The current genetics discourse compels members of the public and medical profession to give serious consideration to a 'natural' desire for healthy descendants. As new forms of prenatal genetic technologies increase, pregnant women's bodies will be the major recipients of these procedures. How do all of these prenatal technologies affect the pregnant body, the major target of prenatal technologies? In this chapter, I also look at the effects of a mechanistic view of the body in reproductive genetics and how reproductive limits are practised on pregnant bodies through a feminised regime of reproductive asceticism.

Besides the discourse of disabled bodies, there is another discourse into which pregnant women are further captured: the discourse on shame. In Chapter 5, 'Gendered Bodies, the Discourse of Shame and "Disablism"', we see how divisions between good and bad pregnant bodies can be made on the basis of their genetic capital and reproductive potential. The social effects and limitations of reproductive genetics in relation to disability as a cultural representation of impaired bodies is illustrated. The techniques of reproductive genetics can have discriminatory effects on disabled people and evidence the promotion of perfection in society. Society attempts to place disabled bodies in social spaces marked by isolation, separation and exclusion.

In Chapter 6, ' Synchronizing Pregnant Bodies and Marking Reproductive Time: Comparing Expert Claims in Greece, the Netherlands, England and Finland', I make cross-cultural comparisons of experts' claims from the four countries in the European experts' study. I attempt to demonstrate that the mechanistic, pregnant body constructed by reproductive genetics is marked not only by definitions of gender and disability but also shaped by space and time. I look at how the disciplinary practices of reproductive genetics are produced in spaces, distinguished by over-arching cultural themes. While the institution of reproduction is considered both spatially and temporally, the questions, 'how and what reproductive space and time are available to women?' and 'how are these culturally constituted?' are asked. I will draw upon the claims of European experts and demonstrate how these claims mark reproductive time for pregnant women and how this time is set within reproductive genetics.

In the concluding chapter, 'Reproductive Genetics and the Need for Embodied Ethics', I offer some brief reflections on my experiences doing the experts' study. Also, I look at the ways experts view ethics and contend that their ethics are disembodied. I contrast these ethics to embodied ethics. I contend that we need to revision women's experiences within reproductive genetics.

1 Prenatal politics and 'normal patient families'

> The invention of the telescope, of techniques used for the cooking, canning, bottling or preserving of the apple or of medicaments to alleviate the stomach pains we will get if we eat too many of them, are seen as technologies – they do not add to our understanding of the working of the laws of nature, but they add to our control over the world around us.
>
> Hilary Rose and Steven Rose, *Science and Society*, p. 1

Introduction

Crick, Wilkins and Watson's discovery of the genetic code of DNA in the 1950s enabled scientists to reveal the chemical dictionary out of which messages serving as blueprints for living structures could be made. The discovery of the elucidation of DNA signalled a turning point in genetics, viewed traditionally as a branch of biology that deals with heredity and the variation of organisms. Breeding by selection, the focus of traditional genetics and of Gregor Mendel's early experiments in the 1860s, was gradually displaced by 'a new genetics' (Cranor 1994). The expansion of this new genetics meant that scientists were able to make biochemical alterations of the actual DNA in cells so as to produce novel, self-reproducing organisms. More importantly, the new genetics privileged the processes of genetic engineering (Minden 1987) which eventually came to be defined as the basis of a novel biotechnology (Bud 1993). The ultimate result was that scientists introduced human choice and design criteria into the construction and combination of genes.

Today, lay people as well as scientists are witnessing the massive proliferation of genetic technologies into many areas of modern social life. Our cultures have become increasingly dependent upon technical advancements in molecular biology. Through these advancements, members of the medical profession attempt to manage, if not prevent genetics diseases or those in which genetic factors play a part. Biomedicine, the medical arm of genetics, is a dominant paradigm in the Western world and in relation to the state, it holds a status analogous to that of the established Church in the medieval period (Currer and Stacey 1991: 1). Genetics occupies a central place in people's consciousness, as biology becomes increasingly the filter through which humans are expected to interpret the world (Lundin 1997). We are continually being told by experts that research into the human genome will lead to immense progress in knowledge, prevention and treatment of disease. On a global scale, medical scientists say that the use of genetic technologies in developing countries will contribute to the prevention of genetic diseases (Spallone 1989). Nevertheless, a growing number of individuals may be or have become labelled on the basis of their genetic information. They face the risk of genetic discrimination. For pregnant women, developments in reproductive genetics are shaping new values for the standards of reproduction – values to which all pregnant women are told they should conform.

Indeed, the medical organisation of childbirth and childbearing has changed dramatically over the past fifty years through the development of prenatal genetic technologies. These technologies are used for foetal analysis, and can be either non-DNA-based (i.e. unrelated to genetics, such as ultrasound scanning) or DNA-based (i.e. related to genetics and blood or serum collection, such as chorionic villus screening, maternal serum screening or amniocentesis). The primary focus in this book is on DNA-based prenatal techniques. However, both non-DNA and DNA-based practices are used in conjunction with each other in the search for foetal abnormalities. Some experts see non-DNA-based procedures as genetic technologies because these tend to be linked procedurally in reproductive medicine. On another level, prenatal technologies are ethically the most difficult applications of genetics (Henn 2000).

In this chapter, I will first focus on the working of prenatal politics and then I will look at the various techniques and procedures used in reproductive genetics. Lastly, I will offer a case study of one of these techniques, molecular genetic testing, and the targets of these tests, 'normal patient families'.

Prenatal politics, gendered bodies and commodification

The English word 'prenatal' comes from the Latin words *prae* and *natalis*, meaning 'before' and 'to be born' respectively (*Concise Oxford Dictionary* 1995). Thus, 'prenatal' connotes the time existing before being born and it appears as if the word itself signifies the foetus (who is 'before being born') more than the pregnant body that carries the foetal body to term. This meaning of 'prenatal' is semiotically loaded. As they concentrate more on the foetus and its health than the pregnant woman, some medical experts working within the discipline of reproductive medicine take this meaning to heart. Experts argue that 'a multidisciplinary approach to the foetus is an essential part of antenatal screening' (Malone 1996: 157). This view suggests that the foetus more than a pregnant woman is the physician's main focus during the prenatal period. The workings of reproductive genetics expose the long-standing feminist unease that the medicalisation of reproduction, pregnancy and childbirth has more often than not been against the interests of pregnant women, making them objects of medical care rather than subjects with agency and rational decision-making powers.

Susan Bordo (1993a: 88) contends that the ideology of the woman-as-foetal-incubator pervades women's experience of pregnancy. Pregnant women are neither subjects nor treated as such, while their foetuses become 'super subjects' (i.e. more important than pregnant women 'subjects'). This representation of women as objects of mechanical surveillance rather than active recipients of prenatal care is an obvious message of pictures displaying the first ultrasound device used in Glasgow, Scotland, as Oakley (1984: 159) demonstrates. But, many prenatal technologies objectify women and uphold this ideology of woman-as-foetal-incubator.

Prenatal politics exist. They are the application of specific ideological beliefs, knowledge and medical procedures on developing (in pregnant bodies) foetuses, viewed as the nascent embodiments of the future. While prenatal politics are more foetus-directed than pregnant women-directed, pregnant women bear the brunt of damaging beliefs and painful procedures. Pregnant women are more done to than the doers, as their foetuses' performances are appraised over time through various technical procedures. Prenatal politics operate in the discursive spaces of knowledge and practices generated by the universalising system of reproductive genetics during pregnancy. DNA, reproductive material, foetuses, gendered bodies and reproductive functions are surveyed and managed in a multiplicity of ways with the effect that pregnant women are compelled to take 'security measures' (Hubbard 1986) necessary for 'successful' reproduction. When pregnant women choose what is generally seen by physicians as the 'correct' prenatal behaviour, these choices may be constructed more by the power conferred on physicians by these technologies and physicians' 'right to choose' selective abortion than by their own pregnancy experiences (Doyal 1995: 141).

Through the workings of prenatal politics, biomedical discourses transform women's wombs into highly managed social spaces – sites of discourses about 'good' genes, women-as-foetal-incubators, 'good enough' foetal bodies and disability. Within the context of the abortion debate in the USA, Nathan Stormer (2000) demonstrates how discursive practices on abortion bring together the womb and the public as a coincident location, prenatal space. For him, the convergence of prenatal bodies and public, political bodies has been accomplished through the convergence of biological and social domains in the discourse on abortion. Within the discourse of reproductive genetics, a divergence not convergence of pregnant bodies and foetal bodies occurs within the discourse of reproductive genetics. The 'producer' (the pregnant woman) and the 'product' (foetus) are detached prenatally by the use of prenatal technologies.

Prenatal politics are generated when pregnant women consume reproductive genetics for the foetus, the reproductive product and attempt to gain knowledge of its quality. In effect, the medical workforce facilitates the commodification of reproduction through the use of prenatal technologies that impart knowledge about the status of foetuses. On the one hand, the concepts such as high risk/low risk, afflicted/non-afflicted and carrier/non-carrier are traditional diagnostic categories, underpinning women's reproductive behaviour and choices. On the other hand, through the commodification of reproduction these same concepts are constructed as descriptions of 'embodied foetuses' with economic labels. These 'embodied' descriptions conjure up various types of foetal body images in the minds of pregnant women, their significant others (partners, families, etc.), medical experts and society. Low risk, non-afflicted and non-carrier foetal bodies are viewed as more valuable both economically and physically in terms of what these potential social bodies can produce and how they are able to contribute to society. Economic relationships are introduced into human reproduction (Overall 1987: 49) because defective foetuses are viewed as prospective, burdensome human beings with a price tag on their heads as well as defective products. Normal foetuses are represented as potential human beings, productive products with a future full of prolific energy. Generally, reproductive technologies evidence a capital-intensive approach to medicine, treating reproductive care as well as reproduction as commodities. Thus, in a context where gender inequality is already present, the negative effects of these technologies upon women, especially the less privileged should not be surprising (Gimenez 1991: 335–6).

Given that the 'products' of women's reproductive activities (conception, pregnancy and birth) can be ranked according to this system of child quality control, women themselves are ranked as 'good' or 'bad' reproducers. Undeniably, we have experienced a reproductive revolution – this technological upheaval in which a diverse series of medical advances have been allowed to insinuate and spread biomedical values (about 'good' genes, disability, women as foetal incubators and 'high quality' bodies, etc.) more indirectly (Lee and Morgan 1989: 3). While biomedicine has a tendency to 'fracture social experience' particularly those of pregnant women (Annandale and Clark 1996), reproductive genetics, emerging from this self-same biomedicine, may also rupture pregnant women's experiences in a far-reaching way. Prenatal technologies have clear social dimensions and values, upholding a reproductive morality. For pregnant women, this usually means that they are drawn into a moral discourse about good foetuses and bad foetuses as well as their good or bad reproducing bodies.

Nevertheless, the technologies of reproductive genetics may have the potential for great benefits. These create possibilities for medical advances and opportunities to make choices about the health of future generations. But, these technologies are value laden and experts are making moral verdicts about foetuses. When looking at genetic technologies generally, Nicholas (2001: 46) contends that these technologies are constructing a new moral landscape and culture. They are disrupting long-established social understandings of how the world 'is', the meaning of the family, the place of humans in the biosphere and the role and responsibilities of the authorised knowledge makers of western culture.

From the above, we have seen that in the workings of prenatal politics, the practice of ensuring that healthy babies are being born has been intensified by the embedded practices of prenatal genetic techniques. Here, these techniques and their development must be seen as more than techniques. Prenatal technologies are the extension of human competence, proficiency, skills and values into the practice of reproductive genetics. In this context, assessment of techniques is the most popular and usual way of evaluating medical technologies (Morgall 1993). This method, 'technology assessment', may be limited as well as somewhat flawed; suggests a one dimensional or too simplistic view of medical technologies; side steps prevention by having no need to search for causes of any given deficiency and avoids important ethical questions. Here, I would contend that over and above technology assessments of prenatal genetic technologies, one clear, if not valuable assessment is that these prenatal technologies have consequences on the workings of social relations that go beyond their immediate application.

The techniques of reproductive genetics

Traditionally, pregnant women underwent prenatal screening for Rhesus factor, HIV, diabetes, etc. However, since the late 1950s, prenatal screening and diagnosis of pregnant women for the detection of foetal abnormalities has increased dramatically (Farrant 1985; Reid 1991; Rothenberg and Thomson 1994). This increase has been aided by the development of new medical technologies, specifically technologies of the new genetics, which aim at the avoidance of common genetic diseases (Wetherall 1991). One clinical geneticist, emphasising that these technologies were in diffusion, wanted all pregnant women to come under the 'genetics' umbrella. She said:

> Genetic screening is well established in antenatal care and I think there could be a highly cost-effective package of making available a genetic counselling contact with all women (sic) as soon as they have a pregnancy contact ...
>
> (E 4)

Prenatal screening

Prenatal screening is the programmatic search for foetal abnormalities such as congenital malformations, chromosomal disorders, neural tube defects and genetic conditions among the asymptomatic population of pregnant women. A general belief upheld in the medical profession is that pregnant women are not forced to undergo screening, as one medical geneticist said:

> If we talk about [the] prenatal screening program in the population, [they have] ... their free will. I mean [there's] not any law saying ... [it is] ... good to take the test [or you must take the test].
>
> (Medical geneticist G 5)

Nevertheless, prenatal screening can cause problems for members of the medical profession. This is mainly because these screenings can be interpreted to mean mass population screenings rather than individual medical assessments of a woman's pregnancy. One public health specialist reported that various interpretations of what prenatal screening means caused debate amongst medical professionals in her country.

> For a number of years, there have been major anxieties about explosions of screening activity and the lack of control of it … particularly as these are public health programs. It's not individual investigation is it? You're screening. Some people … say we should be offering [Down's syndrome screening] … as part of our assessment of pregnancy rather than seeing this as a program … The rational for a screening program is that you go to normally healthy people and carry out something which if you were asked to do [in] a straight forward assessment, you might not because for certain groups of women the risk around Down syndrome is very low. Would you offer that as part of the routine assessment of the problem [or] … would [you] do it as part of a population program? And so I think there is a debate around that.
>
> (E 5)

Prenatal screening can also be viewed as the selection of the proportion of the population of pregnant women who is at increased risk for an abnormal condition. The various methods of prenatal screening include advanced maternal age, biochemical screening for neural tube defects and Down's syndrome, ultrasound and screening for recessive conditions such as haemoglobinopathies (sickle cell disorders, beta thalassaemia major, etc.) that involves genetic tests (Marteau *et al.* 1994). Through prenatal screening, women in high-risk groups are identified for additional testing.

Age

Advanced maternal age is currently seen as a risk factor, given that it is estimated that a higher percentage of women over thirty-five give birth to Down's syndrome babies than women under thirty-five. It is generally accepted in the medical profession that these 'older' pregnant women account for 20 per cent of Down's syndrome births (Committee on Obstetric Practice 1994) and that Down's syndrome is the commonest single cause of severe mental retardation in children (Chard 1996).

Triple test

The most usual form of maternal serum screening is the triple test or triple screen. The triple test is a blood test that examines the level of alpha-fetoprotein or AFP (a protein produced by the foetus) and two pregnancy hormones estriol and human chorionic gonadotrophin (HCG). Blood is taken from the pregnant women between the sixteenth and eighteenth week of pregnancy. Laboratory tests measure the levels of these proteins. High levels of alpha-fetoprotein may indicate the presence of a neural tube defect, such as spina bifida or anencephaly. Low levels of alpha-fetoprotein and estriol combined with high levels of HCG may indicate Down's syndrome (Reid 1991). In recent years, Wald *et al.* (1996) used another marker, Inhibin A, which is a placental product and can be elevated in the serum of women with Down's syndrome pregnancies. Wald and his colleagues contended that the detection rate for Down's syndrome is higher using four markers and age than the traditional three markers and age. Here, it should be noted that abnormal levels of these proteins or hormones are not diagnostic themselves. When abnormal levels of these proteins are found, the test merely indicates a potential risk and further tests such as ultrasound and amniocentesis are indicated. A pregnant woman is given a risk calculation based on analysis of the serum in her blood. For example, she may have a 1 in 250 risk of having a Down's syndrome baby. Usually, prenatal services have a 'fixed' cut off figure (such as 1 in 250) at which point physicians will prescribe prenatal diagnosis if risk calculations fall below the cut-off point.

One expert, an obstetrician, implied that good screenings meant that you would successfully detect Down's syndrome with the further test of amniocentesis. He emphasised that serum screenings were not going to give pregnant women a definite diagnosis of abnormalities and women needed to be aware of this fact. It is interesting also how he uses the word 'test', implying diagnosis rather than screening. He said:

> The idea of [serum] screening test [sic] is to have 5,000 amniocentesis and all are Down's Syndrome ... But if you do that and if you perform the [serum] screening ... you have to inform ... the patient properly because when a woman ... [does] a test (sic), she believes that ... the test (sic) is going to tell her that everything is fine ... The baby is not affected. He (sic) is not malformed. There is no Down's. There are no chromosomal abnormalities but it's not so ...
>
> (Obstetrician G 3)

Another expert noted that serum screening does not allow physicians to promise anything to pregnant women, but it can indicate other 'things' (diseases, conditions, etc.) such as nephrosis.

> Well, the serum screening doesn't really detect anything else, so we can't promise. Of course high AFP detects some other things ... [For example] congenital nephrosis which is another important thing in the high AFP group, because you never can see anything in ultrasound with that disease so the AFP is the only way to screen that.
>
> (Clinical geneticist F 16)

One obstetrician believes that those who are against prenatal screening are involved in a power struggle in which obstetricians and gynaecologists are winning and geneticists losing:

> People who are against screening in general and [the] triple test in particular are against it because ... it is a power issue ... between gynaecologists on the one side and geneticists on the other side ... [Screening] tends to take away from the geneticists' prenatal activities as far as prenatal testing is concerned ... and brings it to the gynaecologists [and obstetricians]. Screening is something that is very familiar to obstetricians ... because they do it all the time. They ... do risk calculations all [of] the time. They take blood pressure ... They calculate the risk of developing anaemia ... They measure haemoglobin ... They act upon screening results. This [serum screening] ... tends to take this activity out of the hands of the geneticists into the hands of obstetricians. This is a power issue ... between two professional groups. Yes, because the geneticists from the beginning, say 20 years ago, they had a say for example, where maternal age limit has to be put – 40 years or 38 years.
>
> (NL 2)

Diagnostic ultrasound

Diagnostic ultrasound is often used routinely in pregnancy as an all-purpose guide to foetal development. It is a procedure that forms an image of the foetus (i.e. foetal imaging) by using sound waves. A video image called a sonogram is displayed on a monitor as the sound waves can be converted into an image. Physical features of the foetus can be seen and often a photo can be produced from the monitor image for prospective mothers or parents. It is common belief that ultrasound hastens maternal bonding (Green 1990a). Ultrasound scans can help to diagnose a number of birth defects including hydrocephalus, limb or organ deformities, some heart defects (Buskens *et al*. 1995), neural tube defects such as spina bifida and renal tract anomalies (Malone 1996). Skupski *et. al* (1994) argue that while ultrasonagraphy never will be perfect in the detection of anomalies, it is viewed as a cost effective procedure.

Prenatal screening using molecular genetic tests

Pregnant women can also be screened for recessive conditions such as haemoglobinopathies (sickle cell disorders, beta thalassaemia major, etc.) or Tay-Sachs disease that involves molecular genetic tests. Analysis from blood samples, CVS or amniocentesis is used where a disease causing mutation has been identified in a family.

Prenatal diagnosis

Prenatal diagnosis is undertaken to determine whether a pregnant woman with a foetus considered to be at risk of being abnormal by prenatal screening process does, in fact, carry a foetus with the disorder in question. Prenatal diagnosis is the identification of an abnormal condition in the foetus (Beekhuis 1993). The methods of prenatal diagnosis include second trimester ultrasound screening, amniocentesis or chorionic villus sampling (CVS). In biomedical terms, the primary aim of prenatal diagnosis is 'to provide an accurate diagnosis that will allow the widest possible range of informed choice to those at increased risk of having children with genetic disorders, within the boundaries established by society' (Advisory Committee on Genetic Testing 2000).

Second trimester ultrasound

Direct examination of the foetus can be carried out by second trimester ultrasound, which can discover developmental lesions and major congenital malformations. High-resolution equipment can accurately assess foetal development and anatomy at 18–20 weeks of gestation.

Amniocentesis

Amniocentesis may be performed in the third or second trimester. Amniotic fluid is drawn from the amniotic sac around the foetus with a long needle through the pregnant woman's stomach. Because the fluid contains foetal cells, it is used to obtain genetic knowledge about the foetus. Amniocentesis can detect Down's syndrome, blood type, metabolic problems (i.e. Tay-Sachs disease) and neural problems. The results of this test are available within seven to ten days. One clinical geneticist indicated that pregnant women might feel that they know enough about these procedures but when things go wrong, difficulties arise because pregnant women 'want the results both ways'. He implies that they want both knowledge of the procedures and good results. He says:

> People do learn what they need to. That's why I was saying that a person's [sic] perception about whether they know enough to make a decision on the Down's screening [i.e. amniocentesis], may be very different to a doctor or midwife's view. They [doctors, etc.] might think this person [sic] doesn't know enough, but this person [sic] feels they do know enough. When things go wrong, then of course [it's] 'Why didn't you tell me so?' ... They do want it both ways; there is no doubt about that.
>
> (E 3)

Chorionic villus sampling

Chorionic villus sampling (CVS) is another way to look at foetal chromosomes and it is performed at the tenth to twelfth week of pregnancy. In this procedure, physicians remove a tiny sample of chorionic tissue with a small tube that is inserted into the vagina through the cervix to collect (with suction) a tiny sample at the edge of the placenta. The sample can also be taken in a similar way to amniocentesis. Karotyping, the arrangement of chromosome pictures in a standardised way, is prepared from the tissue sample. Results are available in seven days. Unlike amniocentesis, CVS is unable to detect neural tube defects and the risk of miscarriage is slightly higher than amniocentesis.

For both amniocentesis and CVS, there is a risk for miscarriage and foetal loss rate at 0.5–1.0 per cent and 1–3 per cent respectively (Advisory Committee on Genetic Testing 2000). In biomedical terms, it has been argued that invasive genetic testing are the gold standard in foetal diagnosis (Kuller and Laifer 1995). Some (Heckerling *et al.* 1994) contend that pregnant women's choice of amniocenteisis or CVS is a utility driven decision which means that it functions to ensure the birth of a normal, not disabled or sick, baby. This obviously has effects on those who are born with genetic defects or chromosomal disorders.

One expert, a researcher, felt that prenatal technologies caused genetic discrimination and she was unhappy about that fact:

> One of the prime areas where you see expansion in genetic diagnostics occur is ... in antenatal and prenatal screening. What are the implications or the main information on the human reproductive processes? The next generation of concerns ... are already coming to the fore. These ... not only touch on women's immediate lives and clinics' but everyone's and concerns things like genetic discrimination ...
>
> (E 1)

Current everyday practice in maternity health care now includes the above mentioned technologies for prenatal screening and diagnosis. While these are the main technologies, others include foetal blood sampling (e.g. obtaining foetal blood from the umbilical cord) and foetal tissue sampling. (When no other tests are available, sampling of foetal tissues is possible.) This practice of looking for foetal health will be intensified with the potential development of newer prenatal genetic techniques such as analysis of foetal cells in maternal blood, a potential technology predicted to be available in the near future (Al Mufti *et al.* 1999).

Indeed in most Western countries, it is quite commonplace that if a malformed or abnormal foetus is detected, the pregnant woman is told this knowledge and she is offered the option of a selective abortion (Asch *et al.*1996) which is an abortion for medical (i.e. congenital or genetic) reasons. One geneticist noted that prenatal procedures may imply action but it also means that one's genes are no longer personal. He said:

> When genetic screening is used for prenatal diagnostics or for detection of any disease, it always has ... relevancy for action [i.e. selective abortion]. You can do something ... In this case, somebody else might take the decision for your genes ... It is no more your personal thing.
>
> (Geneticist F 11)

Prenatal screening and diagnosis of foetuses allows physicians to perform selective abortion of affected, abnormal foetuses. As noted earlier, any specific prenatal technique appears as one of many solutions to the problems that modern reproductive medicine seeks to solve. For example, physicians, especially obstetricians, want that their patients, pregnant women, to deliver healthy babies. In their view, the above mentioned procedures are seen as techniques which allow this to happen. This is regardless of the fact that if any of these required screenings or tests are found to be positive, this may cause severe anxiety and psychological stress for pregnant women (Santalahti 1998). Their abortions may be painful and it is known that some pregnant women do experience distressing consequences in their experience of these techniques (Tymstra 1991; Katz Rothman 1994, 1996, 1998a, 1998b; Markens *et al.* 1999).

If women are autonomous in the process of prenatal screening, these screenings should support or at least facilitate their reproductive rights (Gregg 1995). In this area, women have internalized given social expectations concerning the physical condition of their pregnancies (Green 1990b). But, if women are not autonomous within prenatal screening procedures, they may experience themselves as being used as foetal containers (or incubators) or have the feeling that theirs is only a provisional pregnancy until the 'foetal quality control investigations' have been registered as acceptable (Clarke 1997b: 124).

In this context, one expert notes that what he calls 'risk reproductions' are not viewed as something that can be prevented. This implies that even with all the screening programmes in the world, medical personnel are unable eliminate 'risk reproductions'. This medical geneticist says:

> Those screening programmes were risk assessments and choices [made] for eventual pregnancy termination ... Risk reproductions are not seen as a possible [thing] ... to prevent.
>
> (Medical geneticist NL 4)

In the next section of this chapter, I will look at the implications of one of these prenatal procedures, molecular genetic testing and the effects of these tests on 'normal patient families'. I want to examine some of the claims experts make about this procedure as well as the impact of their claims. What is of interest here is that when reproductive choices are being made, experts' concerns tend to concentrate on the normative, heterosexual family rather than the potential pregnant woman.

Reproductive genetics: applying techniques on 'normal patient families'

When discussing molecular genetic tests, many experts (n = 34 or 74 per cent) discussed in some detail, either 'families' or 'families at risk'.

A shared perception concerning the sorts of families deemed appropriate to treat 'genetically' emerged in their accounts. This perception involved distinguishing between patients who were 'normal' (i.e. without genetic problems and therefore outside the domain of genetics) and those who were 'normal patients' (i.e. families with a genetics problem). For example, a clinical geneticist stated:

> Most people working in this field see a distinction here [i.e. between families] ... We don't see the sort of normal people ... We see the normal, patient family ... a family with a genetic problem ... I think [they] will always be there, they will always need genetic counselling and they will want genetic testing for a defined problem that is known in that family.
>
> (F 17)

The above statement confirms Richards's contention that patients are 'now families, not individuals' (1993: 578). In this context, the Genetic Interest Group (1995) concluded that genetics was distinct from other fields in biomedicine because the component of concern was the family. Thus, the effect of seeing a geneticist is a family issue with implications for several members, as the following excerpt illustrates:

> [Genetic data] are shared by your siblings, parents, sisters and brothers ... What [they] find in you does not just concern you, it concerns the whole family. It makes the situation much more different than if it was just your own.
>
> (Clinical geneticist F 4)

As noted in an earlier discussion, powerful concepts such as 'risk', 'affected offspring', 'viability', 'defective genes' and 'carrier status', along with all sorts of technologies (e.g. amniocentesis, maternal serum screening, CVS, genetic testing) are mobilised by experts in describing 'bodies in families' with genetic problems. When a family undergoes genetic testing, this experience tends to dominate the whole course of a pregnancy. The terms, 'carrier status' and 'risk', take on new meanings, as families attempt to balance the relationship between how experts perceive their levels of 'risk' and the family's own perceived levels of 'risk' (Parsons and Atkinson 1993). One consequence is the production of a repertoire of risks through which families confront as well as manage their reproductive behaviour. In this process, experts mobilise an individualistic, mechanistic view of gendered bodies, effecting reproductive behaviour in families. The full complexity and significance of any family's reproductive history can be lost, as families attempt to structure genetic knowledge in a meaningful way. Experts' mobilising of their powerful conceptual armamentarium makes families' gathering of genetic knowledge a complicated process, protracted over time.

One expert discussed a particular couple who perceived themselves at risk of an inherited genetic disorder. It was only after two pregnancies that their carrier status and genetic risk became clinical knowledge – knowledge that obviously affected their future reproductive choices. But, gaining definite knowledge involved the expert and this family building a picture of the particular couple's risk status. This was an erratic process, extending over three pregnancies.

> We had a couple … The mother's cousin has a child with (genetic disease) and they wanted a gene test … She was found to be a carrier … During pregnancy, [she had] a spontaneous abortion and they forgot about it … When we got the result, we … asked after the father's sample … It had not been taken … We got the … sample … He was found to be a carrier too. So there was a risk in that family. They had a healthy child earlier. [After these pregnancies] they had another … which was an affected foetus … That was terminated also.
> (Clinical geneticist F 16)

One medical geneticist claimed that families wanted a sense of certainty about their risk status. He contended that for families to discover their level of risk was for them experienced as 'a relief'. In this context, Richards (1997) argues that the need for certainty about risk status is part of a general mentality in families, a mentality founded on a 'lay knowledge of inheritance'. This knowledge is based on a molecular model, deeply rooted in contemporary biomedicine (Henderson and Maguire 2000). Within this lay knowledge, families find relief in experts' assessments of their risk status. However, one expert claimed that assessing risk status may satisfy a family's need for certainty (i.e. 'to be sure') regardless of the result of their risk assessment (whether the risk is high or low):

> It is a relief for the family if they know that they can have a prenatal test … If you tell them that you have a Down's child but … you are young … and have a very small chance of having another. They ask: 'How high is our risk?' Let's say that it's one percent … There is another family with a recessive disease with a

high risk of twenty-five percent. My experience is that where a family already has an abnormal child – whether it is a one percent or ... twenty-five percent [risk] does [not] make too much difference. They want to be sure about their next child. This is ... the mentality of the families. They want to be sure [and it] does not make too much difference if it is one per cent ... five per cent or twenty-five per cent. This is my experience.

> (Medical geneticist G 5)

One expert noted that side by side a family's quest for certainty is the notion that knowledge on their 'gene level' is more often than not uncertain. As families attempt to structure meaningful genetic knowledge, they depend on expert certainty. The irony is that experts may not 'really know':

There has been this huge development in genetics, so that we have been able to screen all possible things ... Of course, the knowledge is much more accurate, exact and certain. But still ... you don't really know if you look at your family what you are going to expect ... even if you have the knowledge on the gene level. There are very few diseases about which you really know what is going to happen.

> (Lawyer F 15)

Regardless of the perceived tension between uncertainty and certainty, some experts believe that most if not all families are 'eager' to find out their genetic make up. Finding their genetic make-up was viewed as an opportunity 'to get to know what was possible' for them as a family. Therefore, 'knowing their genetic makeup' was perceived as having a certain status in the repertoire of a family's reproductive choices. Experts claimed that at the very least knowing enabled families to identify the disease in their families and whom it affects. Thus, a clinical geneticist said:

They want to know what it means to the person who is affected, what sort of treatment can be offered and what are the causes of the disease ... [They want to know] anyone in the family who it concerns. So it can be ... a young couple with a child who has ... a genetic disorder and they want to know what all this means ... If it's their first child, [they] want to know whether this will affect subsequent children ... and what the risks are ... So I find myself in a sort of balancing [act] ... in what I say to people ...

> (F 17)

This 'balancing act' was all about assessing how much genetic knowledge experts needed to provide if families wanted to be more certain of their possibilities 'as a family'. She continues:

> Families ... are eager to find out what is possible ... I know a family who had one son with a ... severe genetic disorder and [he] died – the first child ... They are expecting a baby and they [want to] make sure that the boy who died had this disorder ... They are now expecting a girl. So ... though the boy had this disorder ... the girl will not be at risk. And the situation with this test is that the gene has been found ... But not all have been identified and sequenced ... Part of the gene can be ripped out for screening for mutations. If this can be done for this family, this will be a useful result. If they find a mutation in one of the known parts of the gene, the family will know.
>
> (F 17)

In a related context, a gynaecologist claimed that for some families, knowing their genetic make-up could mean that they would not 'take that risk again' (i.e. have an affected child):

> Usually if the family ... knows what it is to have a disabled child ... families do not want to take that risk again.
>
> (F 9)

The above comment suggests that some experts perceive families' need to know their genetic make-up as somehow empowering, if knowledge of risk enables clearer reproductive choices. 'Knowing their genetic make-up' implies that families have the resources to make important reproductive choices which they, the experts, would help to facilitate. One expert asked simply:

> How do we actually use [our] knowledge in terms of empowering people?
>
> (Medical ethicist E 7)

At the same time, experts' perceptions of 'families' need to know' tended to lend support to cultural idioms, if not misconceptions: 'true' kinship is genetic and a proper family is a domestic unit grounded in blood ties (Shore 1992: 300) or a 'proper' family is linked by a blood tie which, in turn, is equated with a genetic tie (Strathern 1997: 48).

Experts claim that if a family is unable to fulfil 'true kinship' because of what is identified as 'bad' genes, the family relies on an expert to ameliorate its situation both now and in the future. In this alleviation process, some experts may gain professional credibility, if not divine status (Katz Rothman 1998b: 502). Yet, when experts detect 'bad genes', something 'wrong' or a defective 'child', this has wide implications, going beyond the clinical relationship. Simply, experts implicate most, if not all kinship ties, as one policy maker noted:

> It can be a great misery for the family and ruin their entire lives to bring into the world a child which has a ... big defect ... it's something very important in one's life to get a child with a big defect ...
> The life not only of the child, the family and the parents but of all the family and the other children are affected by a child which is defective.
>
> (G 1)

If the key to reproduction is how 'good' genes mix, what people do to mix genes may appear less important. Nonetheless, 'good' family planning emerges as a visible disciplinary practice, which has been linked by families and experts to genetic assessment. Pregnant women and indeed 'pregnant couples' (Shildrick 1997: 23) tend to accept prenatal screening and diagnosis along with routine prenatal care (Press and Browner 1997), as these prenatal technologies become daily practice. Yet, some pregnant women, couples and families refuse prenatal genetic testing.

More significantly, an already captive, yet 'voluntary' population of families exists under the rubric of older non-controversial medical practices (such as family planning, Planned Parenthood and contraception services). One expert argued that it is through these traditional routes that the need for prenatal genetic testing becomes visible:

> In the hospitals there are services that give information on family planning ... Within this ... a small part is prenatal screening ...
> The idea of family planning [goes back] many years and they ... are giving [genetic] information and counselling.
>
> (Lawyer G 4)

Another lawyer noted how involvement in family planning might lead to asking 'existential' questions, challenging families to consider newer technologies in traditional medical settings:

In family planning we … inform [people] that it [i.e. prenatal genetic testing] is voluntary. We inform them … about the results or possible results. If you go on this program, [we ask] 'What does it mean for you, your children and the rest of your family?'

(F 15)

Linking reproductive genetics with older non-controversial medical practices, such as family planning, serves to fuse the interests of experts, families and the health care system, a strategy necessary for legitimation of further developments. Qureshi and Raeburn (1993) argue that people's contact with genetics, opportunistic or otherwise, facilitates their openness to it both in the short and long term. Placing genetic technologies within the domain of older non-controversial practices may encourage family compliance.

In this context, Jonsen (1996: 9) contends that while gathering of genomic knowledge in a multiplicity of catchment sites will inevitably change the psychology of ordinary family life, it could be argued further that this knowledge is needed equally by 'normal patient families' and experts alike. Maximising choice and genetic survival for an expert's 'patient family' may be one way of maximising beneficence for oneself (Lilford *et al.* 1994) as a valued and successful expert.

We have been looking at the technique level of reproductive genetics in this chapter and we have seen how prenatal politics are embedded in current biomedical practices in reproductive genetics. Experts as producers of genetic ideology make claims about these techniques and procedures and these claims have an impact on families involved in the institution of reproduction. In the next chapter, the knowledge level of reproductive genetics will be explored.

2 Biomedical knowledge and interests

Genetic storytellers and normative strategies

> Know-how about making or using tools or machines affords a measure of power not only over matter but also over people ...
>
> Cynthia Cockburn, *Machinery of Dominance*, p. 6

Introduction

In this chapter, I want to offer a depiction of the knowledge level of reproductive genetics with special reference to the organisation of knowledge developed around the use of various prenatal techniques, discussed already in the previous chapter. In the first section of this chapter, I will demonstrate that this knowledge level has two inter-related elements: the 'modernist' knowledge base of genetics and the knowledge interests of experts – scientists, geneticists, researchers, physicians, etc. – who are working in this field and making expert claims. Similar to Locke (2001), I reject the rationalised image of science which views science as an 'universalistic, asocial monolith', the appearance of a pure technical understanding unhampered by subjective personal or group interests. Indeed, social scientists have exposed years ago the myth of the neutrality of science (Arditti *et al.* 1980). Following from the ideas of Donna Haraway (1991), who contends that science is about knowledge and power, I would assert that an interesting use of knowledge and power is embedded in the biomedical science of reproductive genetics, as we shall see.

In the latter part of the chapter, I will explore experts as genetic storytellers and review the normative strategies they use in spreading their genetic knowledge. I will demonstrate how knowledge about genetics and in turn, reproductive genetics is developed by a multiplicity of genetic storytellers who have their own knowledge interests concerning the role of genetics in health and society. Knowledge of genetics is so powerful that it alters our conceptions of bodies, societies and ethics (Wertz and Gregg 2000). Here, a key notion informing the ideas presented in this chapter is that all knowledge developments within genetics should have clear and strong links with the field of ethics and these should be embodied ethics – responsible ethics framed by and through our embodied relationships with the others. (The issue of embodied ethics will be discussed later in Chapter 7.)

If, as human subjects, we are grounded in intersubjectivity (i.e. embodied relationships with others), it is through intersubjectivity we 'are capable of hearing and responding to the call of the other' (Martin 1992). In this context, Bauman (1993: 90), reflecting on the 'Other' in the work of Emmanuel Levinas, says: 'I am responsible for attending to the Other's condition; but being responsible in a responsible way, being responsible for my responsibility demands that I know what that condition is'. Here, Bauman sees the 'Other' with whom we are in close proximity as our dependence on our moral status as human beings: the 'Other' for who I am. In this sense, I act for the Other's sake because I am a moral person. From a feminist perspective, the 'Other' would most probably be the 'self-in-relation' with the moral mandate to care (Gilligan 1982) or the 'social self' in relationships and a community (Friedman 1992). In this chapter, consideration will be made of how, with a sense of responsibility, we can know better what the Other's condition is; particularly if this condition not only comes under the scrutiny of genetic explanations that involve subtle, gendering processes.

Abby Lippman (1994) argues that through the genetic constructions of prenatal testing, the reassurance a pregnant woman gets from prenatal diagnosis is a 'biomedical fix', which disempowers and increases her dependency on technology. On the one hand, these technologies are gendered, making her relationship to them a gendering process. On the other hand, they are 'disembodying', reconstructing her pregnant body into a breeding appliance. Drawn into these complex processes, pregnant women are ruled by a fragmented morality of the body; their bodies become 'broken'.

Biomedical knowledge of reproductive genetics

In order to understand better biomedical knowledge of genetics and its development, we should look at the impact of postmodern thinking on the philosophy of scientific knowledge. Current postmodern thinking reveals that a critique of Enlightenment thought, termed 'modernity', has become routine or the intellectual order of the day.

Postmodernists have challenged ideologies of the subject as disembodied, male, white and middle class by developing alternative and
different notions of subjectivity (Huyssen 1990). They have abandoned
the 'modernist', Enlightenment knowledge base (Giddens 1991, 1992)
with its 'universal man' (sic) split into mind and body in favour of
'discursive knowledge' (Habermas 1984) with 'a local embodied (and
gendered) subject'.

As a result of the postmodern ways of thinking, the epistemological
base and methods of Western science and technology are continually
being called into question. Postmodern philosophers, theologians,
ethicists and social scientists distance themselves from what has been
called, the 'episteme of domination' (Benhabib 1990). This 'episteme
of domination', viewed as 'modernist', is characterised by the knowledge seeker's quest for control and the imposition of homogeneity on
the world. This quest is enacted through concept building and the use
of technology. The ideological rational of our technological society is
the pursuit for improvement or what Sarah Franklin (1997) aptly
calls 'embodied progress'. In a real sense, networks of knowledge of
genetics construct the future as an area that can be brought under
some kind of control (Davison *et al.* 1994). This search for improvement through the 'technological fix' privileges means over ends or
leads to, what Zygmunt Bauman (1993: 188) calls the 'declaration of
independence of means over ends'. Technological innovations are
privileged above human processes.

Donna Haraway (1991: 8) argues that the making of science and
technology is a collective process in which present-day scientific
experts attempt to legitimate the 'principle of dominance'. She sees
this principle as embedded in the theory and practice of science as well
as directing the content and social function of science. Within this
collective process, experts are identified as those individuals who can
successfully lay claim to specific skills or types of knowledge, which
the lay person does not possess. Experts are promoters of the corrigi-
bility of knowledge and thus, their role is to protect the impartiality of
coded knowledge (Giddens 1994: 84). Through the intellectual standpoint of postmodernism, we are able to uncover some of the ethical
problems and authority imbalances in the modernist discourse upon
which contemporary reproductive genetics is based. Here, three criticisms appear in a postmodern appraisal of the knowledge base of
reproductive genetics.

Firstly, the building of the reproductive genetics knowledge base
may be more about the privileging of scientific control than human
liberation or health. However, as Rayna Rapp (1994) contends, there is

no simple feminist response to the question as to whether reproductive technology is liberatory or socially controlling because it is always potentially both depending upon the weight various social and individual experiences hold in a particular woman's life. Nevertheless, expert control exists, while the techniques of reproductive genetics are integral parts of controlled scientific reproduction (Spallone and Steinberg 1987b: 15). This controlled scientific reproduction fragments the meaning of motherhood (Hill Collins 1999: 279) and brings both the physician and the pregnant woman into a system of normative surveillance (Balsamo 1999).

In this context, some experts, under the name of objectivity and the claim of scientific neutrality, use categories of systematic justification, such as Reason, Truth, Human Nature, History and Tradition. They appear to embrace modernity wholeheartedly. With these objectivist discourses, scientists and experts use their epistemologies as justificatory strategies (Harding 1990) and uphold their knowledge claims. While it has been argued that human genetics is the science of difference (Murray and Livny 1995), genetic experts by seeking knowledge, understanding and control over the human genome through concept building and utilisation of various technologies may unwittingly attempt to impose a certain level of sameness or homogeneity on the world. On the other hand, as Hartouni (1997: 119) suggests:

> Sameness, repetition, and replication ... are not the issue, at least not in the terms in which they have been presented as such. The issue, rather, is how to conventionalise and contain diversity and (the proliferation of) difference(s) or *how to render diversity and difference socially legible*, and that is what geneticizing both would seem in the end, at least ostensibly to accomplish (author's emphasis).

Secondly, the knowledge base of reproductive genetics involves choices, some of which may be viewed as less than good or irresponsible. For example, in contemporary biomedicine we are witnessing what De Gama (1993) calls resistance to autonomous motherhood. This resistance reveals the problematic and politicised nature of human reproduction. Selective abortion of abnormal foetuses may be the choice of some pregnant women, but this choice may be shaped by docility and pain. The correctness associated with accepting these technologies and feelings of normalcy, experienced by women undergoing other types of medical surveillance (Bush 2000), may also be the experience of pregnant women. Simply, pregnant women are normalised in and through the medical discourse on reproductive genetics.

At the heart of the genetics discourse, there is an intellectual marriage between the epistemological concept, the unity of the abstract, universal man (sic) and the scientific quest for a faultless human gene pool. The strength of this intellectual marriage tends to glorify health as human wholeness and envisage bodily perfection as a real possibility. Here, two basic moral principles are shadowed: the social demand for diversity and human beings' moral need for individual difference and autonomy.

Viewed in this way, the development of the knowledge base of reproductive genetics can be assessed as a process of social choices: implying that there may be 'good' (i.e. responsible) as well as 'bad' (i.e. irresponsible) choices that have been made, are being made, or will be made. With an eye on the future, thoughtful experts should recognise this fact and allow for the possibility of saying 'no' to some aspects of these technologies. Simply, this means on a very practical level that experts may begin collectively to restrict future use in some areas. Here, the notion of responsible social choice is given a social value within the reproductive genetics discourse.

Thirdly, the knowledge base of reproductive genetics may involve the genetic discrimination of both the 'medically' and 'socially' 'deficient'. While this is a familiar argument, I would contend that as this knowledge base expands, experts become more able to identify, alter and understand genes and genetic functioning. Experts may also be able to use their techniques and knowledge as the main criterion for making judgements about what genes are 'good' or 'superior' and what ones are 'bad' or 'faulty'. Thus, they may be able to make moral and social judgements not only about the social existence of genes but also about who is fit to be born. In this context, Rapp (1998: 144) contends that prenatal genetic technologies provide a context in which every pregnant woman is interpolated into the role of moral philosopher. Rapp argues further that one cannot confront the issue of quality control of foetuses without wondering whose standards for entry into the human community will prevail.

Here, if experts discriminate between superior and faulty genes what happens to experts' quest for social responsibility? Simply, the effects of the knowledge of reproductive genetics on social and health care values may determine whose genes and not just what genes are to be discriminated against. One problem implicit in this discriminatory process could be that the concept, 'human gene pool', will be replaced by the historical, social question: Whose super 'human gene pool'? Here, the message to a disabled person could be: 'Humans like you are being removed from the human gene pool'. In these processes, there may be little consideration of the views of disabled people and others labelled as not only medically but socially deficient (Morris 1991). The above discussion illustrates how the tradition of scientific or expert disinterestedness (i.e. neutrality), the discourse of objectivism, is perhaps, being replaced by a new ethic: the making of moral judgements by experts. Does this shift evidence a major shift in the paradigm of science in general and biomedical science in particular? Will scientific knowledge, specifically genetic knowledge and techniques, no longer be viewed as neutral by the general public? The answers to these questions are most probably affirmative. Thus, the logical conclusion is that any expert's knowledge base can be used for destructive as well as creative purposes and experts may need more public accountability and scrutiny than exists at present.

Knowledge interests in reproductive genetics

Knowledge of reproductive genetics is developed in networks by a multiplicity of actors who have their own knowledge interests concerning the role of genetics in our bodies, lives and societies. Lay and professional discourses on genetics touch a myriad of lives and effect current definitions of health and illness. In the complex process of building genetic knowledge, there are members of the lay public including both consumers and onlookers. There are the professionals or 'experts' including scientists, physicians, philosophers, ethicists, social scientists, lawyers, judges, police, military personnel, environmentalists, athletes, policy makers, politicians and so on. The intellectual issues they all are concerned with are deeply intertwined with social and political questions concerning the privileging of scientific knowledge and nature over ethical and social values (Haraway 1990). Nevertheless, those with either lay or technical knowledge in genetics must be able to differentiate very clearly between notions of population control and health promotion. For example, quality control of the human population should not be the aspiration of reproductive genetics.

We must first be aware of basic human values within science in order to understand new developments in reproductive genetics. Science must be free and creative – a source of liberation for humankind – while at the same time, those working in science must obey rules and principles which protect embodied human beings. But, we know that science has not always been managed well. For example, if we reflect on scientific policies that were upheld by German geneticists during the Second World War, human life and a human being's need for individual, autonomy and freedom were not protected (Steinberg 1992).

Similarly if in today's world, prenatal technologies are used to enforce selective breeding practices, such as the elimination of undesirable individuals or the altering of the hereditary qualities of a race or a population, genetics can be confused with eugenics. The doing of genetics must be done by moral and responsible persons. If Bauman (1993: 31) is correct that 'being a moral person means that I am my brother's [sic] keeper' then we must explore the complexities of this ancient idea. In turn, we must find the responsibility within in ourselves – in relationship to the Other – to be our brother's and sister's keepers. In looking at the development of knowledge interests in reproductive genetics, we turn our attention to experts as genetic story tellers who employ a variety of normative strategies in order to uphold their knowledge interests as they do genetics work.

'Genetic storytellers'

Barbara Katz Rothman (1998a: 18) has argued that genetics is not just a science: it is a way of thinking, 'an ideology for our time'. She contends that scientists constantly make decisions about what they consider significant and that their choices reflect the society in which they live and work. For Katz Rothman, experts or scientists are not detached observers of nature: they produce culture. In a similar vein of thought, Lippman (1994: 12) constructs a view of biomedical experts as the primary 'storytellers' – the whole range of professionals that have a 'wealth of raw material from which to choose when they construct their explanations, their stories, for the conditions that interest them'. These storytellers make complex genetic narratives accessible for popular consumption. This is a major part of their 'genetics work', as they attempt to establish their authority by the stories they tell and the metaphors they use (Rosner and Johnson 1995).

In this context, Duster (1990: 18) contends that while experts work within the prism of heritability, how this genetic prism refracts and thereby selects problems for research, screening and treatment is a matter for the sociology of knowledge. He argues that at specific points in time, particular social questions emerge for detailed inspection and set the path for the prevailing epistemology of a culture. In his view, today's path is science under the banner of the new genetics. Duster insists further that while experts say that the new genetics will bring greater health to various groups in society, their approach does not address effectively the 'truth' claims of genetics; claims that are shaped by the workings of contemporary science and the cultural need for genetics. For example, in making truth claims, experts establish 'scientific proof' through particular lines of enquiry or paths of knowledge (Duster 1990: 21). By setting up research programmes and influencing public policy, they legitimate their authority. On the other hand, their claims are shaped by the concerns of a culture in which there is an increased role for the explanatory basis of genetics in an array of behaviours and conditions (Duster 1990: 94), what Lippman (1994), as we have already seen, calls geneticisation.

Here, Duster believes that any scientific advancements, promises or problems, applied to populations will inevitably imply a certain amount of unease, especially for populations at risk. In this context, Higgs (1997: 163) argues that pregnant women, one such 'population at risk', may feel unease, given that many of their own life experiences are being taken over by a detached medical elite. Pregnant women's unease at the direction taken by modern biomedicine can become problematic for them particularly if it is combined with 'manufactured uncertainty' (Giddens 1991): the result of too much information about risks and no way of assessing the impact of these risks. One such example is the controversy over the need for amniocentesis against its potential for causing miscarriages, psychological disturbance or what Katz Rothman (1994) famously terms 'tentative pregnancies'.

The need for normative strategies

How experts theorise about their work in telling genetic stories is illuminating. Here, the focus is on how, in doing genetics work, spreading genetic knowledge and making their claims, experts employ a series of normative strategies in order to be successful genetic storytellers. Regardless of whether they share a consensus, experts employ these strategies in order to construct a theoretical base from their wealth of genetics intelligence. These strategies define what is needed for effective genetic storytelling and a firm knowledge base. These include: (1) claiming ownership of genetic knowledge and practices; (2) maintaining a distinction between the social and scientific; (3) deploying genetic foundationalism; and (4) advocating the application of bioethics. These strategies, shaped by the prism of heritability, ensure, at the very least, a common frame of reference.

Claiming ownership of genetic knowledge

In an attempt to ensure that the science of genetics will persist, experts claim ownership of genetic knowledge and resultant practices, while speculating on the potential health benefits of genetics. One expert expressed his approach to prenatal genetic screening as a 'mass medical activity' for disease prevention. He believes that this 'activity' is the domain of clinical genetics which, he implies, is a distinct discipline within biomedicine.

> Screening is an opportunity for preventing a serious disease ... Here we have a mass screening test ... My approach [to prenatal genetic screening] is basically as a mass medical activity ... This is still very much a specialist area covered by the clinical genetics discipline.
>
> (Epidemiologist E 6)

A clinical geneticist explores this view further. She felt very strongly that only specialists should interpret genetics test results:

> I think ... susceptibility tests in the hands of a geneticist or clinical geneticist are OK but when it spreads out to the whole medical profession, it is not OK. We recommend that all these patients come to the clinical genetics department ... Doctors must know what they are doing when they interpret these tests.
>
> (F 4)

While the human genome may open up broad panoramas of information, ownership is about maintaining the boundaries of genetics, holding genes as medical property to be marketed and bargained over. Kimbrell (1993) says that genes are public material neatly shelved by their proprietors (i.e. experts) in the human body shop. While the course of a genetic disease does not change within these boundaries, the public is given the illusion that a particular diseased gene can change if it undergoes 'therapy'. Paradoxically, one clinical geneticist made it very clear that gene therapy has no 'practical meaning yet' and that the public often misunderstands its current use:

> If we think about knowledge on medical aspects of genetics, I think there are many misconceptions. People often think that gene therapy is ... being done and it is a very good thing ... They have no idea that it is something that is only being tested in some labs ... It has no practical meaning yet.
>
> (F 5)

Still, genetic disease can only be expected, watched or wondered about by both patient and expert. Thus, as Jonsen (1996: 11) contends, the expert quest for understanding genetic diseases and owning the problem focuses attention from a practical to a theoretical, speculative science – from Scientia Activa to Scientia Contemplativa. Jonsen suggests that within reproductive genetics, the expert is compelled to look beyond the presenting patient: the expert is forced to theorise – to consider the relationships, experiences and emotions of the many, linked at the molecular level – both now and in the future. Thus, in order to be successful, experts need ownership claiming as a continuous strategy, affecting both current ideas on their patients' pedigrees and any molecular eventualities.

Maintaining a distinction between the social and scientific

Biomedical experts, similar to other scientists (Latour and Woolgar 1986: 21), distinguish between the 'social' (behavioural processes) and the 'intellectual' (objective science). In their mind's eye, society is constructed as separate from the realm of science. In genetics, this issue has been flagged up by Anne Kerr and her colleagues (see Kerr *et al.* 1997: 295). The reason for this distinction is that some experts may be threatened by the encroachment of external, social factors, particularly politics. Regardless of the fact that some experts are keen to get 'politically' involved to advance the genetics cause, they view their biomedical activities as being driven by science not politics. While the distinction between the social and the scientific tends to be accepted as unproblematic, it may have significant consequences for the type of genetics experts produce. For example, experts' interests in the new genetics are spurred on by their scientific or intellectual interests; concerns that are constantly objectified. As the unchanging course of a disease unfolds, these concerns are 'followed' and 'described'. One expert illustrates this point quite vividly:

> As a geneticist, I have been very much interested in population genetics, learning how to approach populations *intellectually*. So I collected all the patients … in the country … to follow their story and describe the natural history of the disease …
>
> (E 4; my emphasis)

Elucidating this distinction further, another expert discussed what he saw as 'an intrusion from society' by politicians in his area of work. He believed they caused 'political problems' and thwarted his scientific role with their misplaced social interests. This obstetrician stated:

What strikes me ... is ... the involvement of politicians and the ethical problems they see ... They have problems with screening, because ... the end result ... if something is found, ... is an abortion. That's their problem. That's why they don't want to talk about it or when they talk about it, they only do it in negative terms ... aThey always say, 'Oh it's no good because it will cost you a lot of abortions and there will be no advantages because handicapped people are no longer accepted in society' ... They only see the problems ... So the first thing I see in screening is ... the political problems.

(NL 6)

This expert wanted to maintain a clear distinction between his scientific 'genetics' work and the 'social' (e.g. abortions and an interest in disability), perceived as disruptive to this work. In his comment, the 'social' values of politicians do not appear to sit well next to the 'intellectual' ones of the scientist. He wanted to make this mis-match clear. This is an example of how experts are able to draw upon this distinction between the 'social' and the 'scientific' as a powerful resource to characterise their own work. In order to be successful, they utilise this distinction to demonstrate the significance of themselves over and above others and more importantly, the intellectual (i.e. thoughtful, contemplative) nature of their work. This strategy is an elementary part of the experts' repertoire and is consistent with their use of 'gene talk' which, Conrad (1997) argues, functions to over simplify very complex, social issues. As a strategy to maintain professional dominance *vis à vis* others outside of the circle of science, they attempt to ensure the dominance of the intellectual by constantly opposing 'social' intrusions.

The deployment of genetic foundationalism

As part of the above mentioned maintenance work, experts employ a type of genetics foundationalism upon which they are able to ground their scientific endeavours and, more importantly, exploit the cultural significance of their genetics work. McCaughey (1993: 235) contends that 'foundationalism assumes the need for, and existence of, one authoritative framework for distinguishing the right from the wrong, the real from the unreal, and the healthy from the sick'. She argues that it is precisely this universalising tendency that reproduces 'the ability of those who already have privileged access to knowledge production to determine the normative framework and the inability of those with less access to knowledge production to challenge that framework with any credibility' (McCaughey 1993: 236). In this context, eugenics, the concern with the genetic improvement of mankind, becomes fundamental to genetics foundationalism (see Paul 1992: 667).

In discussing eugenics, one lawyer believed that a measure of whether or not genetics is eugenicist is how far it can go. Simply, 'how "preventative" (i.e. of disabilities) is it?' In her view, the more preventative, the more it was eugenicist:

> The more instruments you have in your hands to prevent disability, the less you come to accept it. It's [disability] something that can be avoided. It's preventive medicine but you don't know how far it will go ... You do all of this testing to prevent disabled babies being born with defects ... Will this turn to eugenics or not in the long run? ... How far can preventive health go? Prevention [in] the extreme can become a sort of eugenics if you're preventing things more and more.
>
> (G 7)

In contrast, a clinical geneticist claimed that as a word, eugenics has a bad press and that she wanted to emphasise its 'good side':

> If we take the word, eugenics ... people ... think of the worst things ... the evil, bad things that you can do with genetics. Well it is not [like that]. Eugenics was a good word at that time when it was used for the whole field of human genetics. We wanted then to improve the ... quality of the human race. This side of the word has to be considered more. When we offer these tests, our intentions are to improve the quality of the human race ...
>
> (F 4)

Another defender of eugenics said simply:

> My view is that the principle of abortion for disabling handicaps, for serious disorders that causes disability, is acceptable, however disagreeable that may be.
>
> (Epidemiologist E 6)

In a foundationalist framework, experts express eugenicist concerns. In this way, they attempt to facilitate successful genetic storytelling as well as the productive circulation of a genetics moral order in which one's genes becomes a reproductive resource within the family. Simply, they want to privilege a morality of the body which upholds the standard for conventional (i.e. non-diseased, genetically 'normal') off-spring and society's need for citizens who are fit to be born.

Advocating the application of bioethics

While experts may employ genetics foundationalism as a way of making politics and other extraneous social factors unimportant, if not invisible in their knowledge production process, they need continu-ally to ensure that all relevant normative issues are incorporated into the domain of the intellectual. Indeed, they do this by advocating the application of bioethics, which functions to maintain the significance of the intellectual (i.e. autonomy of science), while at the same time staking a claim in the social (i.e. morality). In practice, this means that if ethical problems arise these can be identified by experts within the context of science and transmitted to the public as accessible, 'social' rather than 'scientific' knowledge. For example, while the medical model of pregnancy may be both 'product orientated' and 'fetocentric' (Katz Rothman 1996: 26), the strategies employed by those upholding this model are utilised to ensure optimum reproductive health outcomes within society (Bobinski 1996: 80). This model and its atten-dant strategies are justified on ethical grounds because health and life are valued over disease and death. In these contexts, reproductive genetics can be seen to rest on firm ethical grounds. As Katz Rothman (1996: 26) says:

> Ethicists who evaluate prenatal diagnosis are often most comfort-able with those situations in which the foetus is diagnosed with an inevitably fatal condition: it might not survive the pregnancy, or even if brought to term and born alive would die shortly there-after. In such circumstances, prenatal diagnosis is generally understood to present no-ethical dilemma. An abortion simply brings the inevitable to a rapid conclusion.

In considering prenatal diagnosis, one ethicist believes that as deci-sions about abortions become more public so also will 'the moral question' (i.e. the ethics of selective abortion). In her view, this will have far-reaching social implications:

The really important issue is what is going to be done after you hear about the results. This is a moral question and I think this is a question of motherhood and [for] each one of us.

(G 9)

Nevertheless, in order to be successful, experts need to ensure that genetics work is done in spite of moral or ethical implications. While experts can be seen to embody paternalism, their authority is deemed worthy of respect and may be appealing for patient and expert alike. The information experts produce may be difficult to refuse or resist. One expert, both a lawyer and ethicist, discussed how easy it could be for some women to be drawn into an expert's mindset, even though they are 'given the choice' to refuse testing:

You know doctors are more paternalistic ... They have a ... traditional role ... Their authority maybe more respected than in other areas ... Women are given the choice. For example, you can ... or if you don't want, you don't [have to] take the tests. But they are given the information in such a way that it is difficult for them to refuse.

(G 7)

Here, in order to be sustained as a viable scientific discourse, reproductive genetics is aided by the moral mediation of experts. Without this, there is a danger that part of the genetics order will break down and expert storytelling will become unsuccessful. Nevertheless, while ethics is a fundamental piece of the knowledge base of genetics and helps to produce this genetics order, ethics is constructed in a piecemeal way. Simply, most if not all experts agree on the need for ethics and are keenly aware of the ethical dimensions of their work. However, their views on the scope, importance and function of ethics tend to vary. Most believe ethics is fundamental in their work; others believe ethics, while important, is relative and still others question whether or not experts could 'hold a view' on ethics (Ettorre 1996). (These expert claims will be discussed further in the final chapter of this book.)

Yet, experts are expected to 'practice', if not embody ethics. Thus, telling genetic stories may mean providing ethical codes for genetic material. Experts fashion powerful judgements about genes, ethics and indeed health through relatively stable ensembles of procedures, instruments, theories, results and products to which they give their allegiance (Wright 1994: 13). In turn, their judgements filter through society and help to construct the view that all adverse consequences of social, psychological, economic and physical functioning flow from not only illness but also genes (Asch and Geller 1996). More significantly, expert judgements can become the engine for further developments in genetics, as ethics and technological advances become inextricably linked.

In this context, one expert suggested that the concern for ethics should go hand in hand with the growth of technology. In a very curious way, his statement was a defence of technological development, privileged over ethics. But, for him it was difficult to consider the 'rights' and 'wrongs' of an issue, when new technologies were being developed at such a fast pace. This policy maker said:

> When science and new technologies develop [it] is sexy, ... attractive, interesting and exciting. Everybody wants to push back the barriers ... instead of saying every now and then, 'Should we be pausing and saying, "Is this something we ought to do?" [and] not, "Is this something we can do?"' The time to say that it is something we ought to do is five years before we can do it ... That's when you need to, before it's in place. Once it's in place, it is very difficult to say, 'No, we ought not to be doing this' ... You need to do it in advance ... The problem is that when the scientists say immediately 'It's available now', it is very difficult to resist it being used ... We have to start saying, 'Now we have limited ability to do ... genetic testing'. But it's going to escalate very rapidly. We need to be making the decisions now about where we put pegs in the sand and where we decide what is right and wrong.
>
> (E 7)

This excerpt illustrates the type of inner logic of science or 'reflexive scientization', highlighted by Beck (1992: 159–60) in his observations on science in the risk society. The portal through which genetic risk is opened up and treated scientifically is a critique of science and its attendant technologies. Based on social definitions and relations, these risks destroy opportunities to resolve mistakes internally (i.e. within science). Scientific expansion and development presuppose self-criticism. Attempts by experts to resolve technological mistakes are transmitted publicly, achieving moral significance. While these publicly transmitted judgements become forced critiques of new developments, a complex legitimating process occurs, creating further technological expansion.

Ethics may appear to function in this context as a check on technology. But, ethics can also be employed strategically as an intellectual bulldozer pressing the frontiers of science. As Katz Rothman (1998a: 36) aptly states, 'bioethics becomes a translator, sometimes an apologist, sometimes an enabler of scientific "progress"'. Experts hope to eliminate genetic abnormalities in offspring. They do this either by biological prevention or apparent genetic cure (i.e. gene therapy). In this quest for healthy babies, experts may experience a moral imperative to push the boundaries of science further (Spallone 1995: 129). Ethics allows them to be prosperous and to do just that.

In this process, Rabinow (1999: 13) contends that through the genetics discourse, ethics, societies and bodies are in the process of reformation. Experts along with capitalism and bureacracy bring together health and identity, wealth and sovereignty and knowledge and values. Furthermore, he argues that the biologicalisation of identity, what he terms, 'biosociality', is becoming embedded in science and understood as inherently manipulable and re-formable. Simply, older biological categories of gender, age and race are being replaced by knowledge of genes – genes that are viewed as determining our prime location of identity. For all of us, these issues become profound questions of epistemology and ontology. For pregnant women, the scientific quest for biosociality is increasingly shaping their reproductive experiences.

I have described the knowledge level of reproductive genetics including both the developing knowledge base as well as knowledge interests of experts. In looking at these knowledge interests, I explored experts as genetic storytellers and their normative strategies, which uphold their genetics work and facilitate the circulation of genetic knowledge. In the following chapter, I will look at the work organisation level of reproductive genetics.

3 Organisation of 'genetics work'

Surveillance medicine and genetic risk identity as a novelty

One sees the wagon dragged back,
The oxen halted,
A man's hair and nose cut off.
Not a good beginning, but a good end
  I Ching (Book of Changes), p. 149

Introduction

As we have seen in the previous chapter, the proliferation of genetic technologies into reproductive medicine involves processes of formation of knowledge networks including medical know-how, knowledge interests, including a series of normative strategies and technical expertise. Genetic knowledge and its development is related to genomics, the study of the genetic control of body functions in health and disease. When genomics is applied in the institution of reproduction, theories on viable and non-viable genes and genetic material are constructed about pregnant and foetal bodies. Authorities on foetal medicine contemplate child enhancement as a real possibility (Parker and Hope 2000). Genomics has profound implications for biomedical research and development as well as the future delivery of health care services (Fears *et al.* 1999). In this sense, genomics, as the theoretical base of the new genetics, can be seen to refine medicine in terms of diagnosing diseases and transforming traditional, clinical practice.

DNA sequence variation has become the hard currency of the new genetics (Spurr *et al.* 1999) and DNA technology, used specifically within reproductive medicine, gives pregnant women more precise information about the risk status of their foetuses (Beech 1995). Indeed, the existing regime of prenatal diagnosis constitutes an important social niche for the entrenchment of a rapidly increasing number of new genetic tests based on DNA diagnosis (Koch and Stermerding

1994). Significantly, within this niche, embodied risks emerge; risks shaped through authoritative medical discourses, the intentional application of prenatal technologies and variable constructions of potential threat inscribed on pregnant bodies.

Biomedicine, as a cultural and social system, is a formation of meaning and behavioural norms, attached to particular social relationships and institutional settings. Reproductive genetics emerges from biomedicine as a form of medical knowledge and practice embedded in the institution of reproduction. Within reproductive genetics, the cultural meaning of experts and their professional norms are not only shaped by their authoritative claims and relations with those they treat but also practised within specific social settings that change. My contention is that the current biomedical discourse on reproductive genetics upholds an observable method of working, facilitating the advancement of sophisticated genetic techniques by a defined group of experts within the regulatory system of reproduction. Biomedical experts mould the institution of reproduction by defining as well as facilitating the organisation of values, norms, activities and social relations around gender, pregnancy and the body *vis-à-vis* human genetics. An idea with stable currency in this process is that the more scientists and experts know about genetic roots of healthy, normal human beings, the better they can predict, treat and correct deviations (Murray and Livny 1995). In this context, two related questions emerge: How is 'genetics work' organised within biomedicine? And what is happening within reproductive medicine to facilitate the growth of genetics? In order to answer these questions, I will describe the method of work that accompanies developments in genetics generally and their consequences. As we shall see, the active administration of genetics work by medical experts is a complex process of embedding genetic techniques into everyday biomedical practice.

Fashioned by knowledge of the growing impact of biotechnology and genetics on reproductive practices, this chapter is divided into three parts. Firstly, I will contend that organisation of genetics in biomedicine is done in and through surveillance medicine. Genetics becomes embedded in the discourse on public health as the site of intervention becomes the community in contrast to the hospital. Secondly, I will highlight the construction of genetic risk identity as novel: this risk identity becomes something which someone 'owns' in an embodied sense. Lastly, I will look why we, as social scientists, should reclaim the genetics agenda and position the body as central in the genetics discourse. We need to assert corporeality in all our work.

Surveillance medicine: organising genetics in the community

Most, if not all, experts in genetics have been trained to look for illnesses or diseases in hospitals or clinical settings. Their work has been organised by the dominance of clinical medicine (Foucault 1973) and informed by an 'old' bio-medical health care model (Bunton and Barrows 1995). Within this model, medicine is characterised as restorative, institutionally based, constructed upon a small group of experts, necessitating heavy investment of resources and concentrated on passive individuals (Bunton and Barrows 1995: 206–7). The paradigm of hospital medicine has been based on separating those who are ill from those who are healthy. The hospital has been the key site for this spacialisation of illness; the space where intervention takes place, where the medical professional manages, treats and deals with disease. Illness and disease have a positive visibility with specific signs and symptoms. Medical professionals search for these indications as they focus on the individual.

At the same time, the emergence of a newer form of medicine based on surveillance of 'normal' populations has developed. Here, the site of intervention is the community in contrast to the hospital. Older techniques of hygiene were translated into newer strategies of health promotion with the dream of surveillance in which everyone is brought into the vision of the benevolent eye of medicine through the medicalisation of everyday life (Armstrong 1995). The ways in which traditional medical science has established 'the space' for illness (i.e. in the hospital or clinic setting) are changing. As a result, the traditional boundaries between health and illness are breaking down. Health is being redefined in terms that assess levels of functioning and well being in everyday living within social contexts (Tarlov 1996). A contemporary belief in biomedicine is that individuals and their ill health cannot be understood solely by looking inside their bodies and brains; one must also look inside their communities, their networks, their workplaces, their families and even the trajectories of their life (Lomas 1998). In this context, the policy challenge focuses a population's attention on preventing disease through public health messages (Blane *et al.* 1996).

Bunton and Barrows (1995: 207) have argued that a new public health care model has emerged and widens the relevant points of social contacts between professionals and patients into social interactions oriented toward the social body. This is in contrast to the old biomedical health care model that relies upon focused interactions between professionals and patients within a clinical setting. In this sense, genetics can be seen to be grounded in this new public health, as it absolves the social structure of responsibility for disease (Wilkinson 1996: 63). Genetics has proceeded on the assumption that genetic knowledge will advance the health and welfare of the community (Petersen 1998). Furthermore, it is within this 'new' space of illness, the community, where healthy or unhealthy life styles as well as the combination of genes (i.e. faulty or not), patient carrier status and risk factors can be identified. In this movement from hospital to surveillance medicine, there has been a shift in modern medical culture from looking at subjective signs and symptoms that people experience when they feel sick to performing objective tests in the community.

This shift has aided the diffusion of genetic technologies, as they are extended into various communities of risk and played out on gendered bodies. Thus, interventions such as prenatal genetic testing affect not only 'normal patient families', as we saw in Chapter 1, but also the general healthy population as well. Pregnancy comes to be seen as risky 'genetically' in popular culture. On the level of prevention, the medical gaze in the new genetics is framed by surveillance medicine and focuses more on the pre-disease, at-risk state and less on the disease itself. Healthy individuals in the community have the potential to become part of an ever growing population of the 'healthy or asymptomatic ill' (Hubbard 1985). They can be designated as being 'sick without being sick' and so also can be their present or future offspring. Depending upon the seriousness of the disease designated and the community's knowledge of genetic variations, stigmatisation or discrimination can be a real concern. Thus, genetic screening is rapidly becoming an issue of public health (Asch *et al.* 1996).

One policy maker referred to the new genetics as a 'public health story', suggesting that genetics may be a new form of intervention in relation to health and illness:

> Biotechnology and genetics … In my opinion, medically speaking, it is a public health story.

(G 10)

Atkin and Ahman (1998: 448) have noted that while the public health message of genetics may emphasise individual decision making over life styles, genetics extends medicine's claims of competence to newer areas of personal and social life, including ideas about appropriate behaviour on the part of patients who are seeking or should be seeking genetic services. Here, one medical geneticist believed that the message of genetics should be heard at an early age – becoming a basic component of health prevention and education:

> People should be informed because we can't have relevant discussions about this topic unless people know what they are talking about ... So I am all for ... advanced information to the public ... They are very perceptive and receptive towards the issue ... genes, diseased genes ... if they receive [an] AIDS information package I think that they equally well should receive a genetic information package.
>
> (F 3)

Additionally for this expert, the appropriate response for a lay person in a public health context was a keen interest in one's 'diseases'. This interest could be elicited from adolescents – those close to 'a fertile age'. The implication is that adolescents are at a ripe age to digest genetic information and should want genetic knowledge, if not services:

> [They should receive] a genetic information package when they are close to the fertile age because ... there is a keen interest then in it. 'What am I?' and 'What diseases are in my family?' ... So I think that would be the right way to reallocate resources from ... medical genetics ... to health education.
>
> (F 3)

Related to embedding genetics within public health, another expert, an obstetrician, notes that a general consensus amongst his colleagues was that appropriate behaviour of an 'aware' thirty-five-year-old pregnant woman was to choose prenatal diagnosis, either amniocentesis or CVS, and the inevitability of this choice. However, he believes that constructing a viable public health discourse, enabling one to make this choice, was indeed 'hard work':

> You have this policy through the mass media, through the newspapers, through the women's magazines ... that ... if you are 35 and more you have a great risk to have a baby with Down's syndrome or chromosomal abnormalities. We tried very hard actually, I can tell you ... We tried very hard to convince ... the patients – the pregnant women – to do amniocentesis or to do CVS for Down's syndrome ...
>
> (G 3)

One expert emphasises the participatory style and preventative focus of the new public health. She believes that people should be clearly informed of their risks as well as new, available, technologies:

> [There] should be health education to inform people about the risk and the existing techniques and make them available to everybody ...
>
> (Ethicist G 9)

As we have seen, with the development of the new genetics, a newer paradigm, surveillance medicine (Armstrong 1995) appears more clearly. The new genetics message is circulated through the new public health and the site of intervention is the community in contrast to the hospital. It is within this 'new' space of illness – 'the community' – where healthy or unhealthy life styles as well as the combination of genes (i.e. faulty or not), a pregnant woman's carrier status and risk factors are identified.

Confirming the community as a viable site for intervention, a clinical geneticist spoke about setting up 'a gene shop' and how this would be an important way of generating interest and public awareness:

> I am involved in ... running the gene shop ... to increase the level of public awareness and ... knowledge ... [So] that when somebody taps somebody on the shoulder and says: 'Would you like a triple test?', they thought that through ... They know ... those that said, 'Yes' have said 'Yes' because they understand why they say 'Yes' a bit more ... It means to have anything to do with genetics ventilated in the widest possible way ... making knowledge freely available.
>
> (E 3)

The shift from looking at subjective signs and symptoms that sufferers, consumers, patients, pregnant women and gendered bodies experience when they feel sick to performing objective tests in the community is a point illustrated by the following clinical geneticist. He discusses the development of a new technology, analysis of foetal cells. He describes how this technique is being developed with the explicit intention of screening large groups of low risk women, previously outside of the scope of reproductive genetics. Stating that this needs to be highly organised in order to be effective, he implies that if analysis of foetal cells becomes rountinised, there will no longer be a need for maternal serum screening, CVS or amniocentesis. A DNA test on foetal cells in the mother's blood is all that would be done: 'no risk', 'no fuss'. But, technically, this is not 'yet' possible. This expert envisages that most if not all pregnancies will come under the domain of large-scale DNA based prenatal screening programmes. If analysis of foetal cells materialised, this technology could spearhead the mass use of genetic tests (i.e. genetic screening) on pregnant women in the community:

> [I] am talking about the possibilities of large group screening of low risk pregnancies for chromosomal abnormalities, by looking at foetal cells in maternal circulation. ... At the moment ... it doesn't work as a test to be performed on a large scale ... If there is a real good programme available for large scale detection ... it's only possible to do it well, if you have a very high level of organisation ...
>
> (NL 4)

Another expert, a gynaecologist, confirmed the possible proliferation of foetal cell analysis in the community and described this as 'magnificent'.

> It would be magnificent if there were a movement towards that ... in the future. Because if it will be a diagnostic procedure, then it will be magnificent ... You don't need amniocentesis or CVS any more ... If it is diagnostic procedure ... it will be too easy for people to do that ... Now they ... have to think about it. What is more important – having an abortion after amniocentesis or the child with Down's syndrome for instance? But, if you are only to take some blood from your arm and you know the result, I think ... it will be too easy for them to have their blood examined and then they know I will get a child with Down's syndrome and now, do I want an abortion for it ...
>
> (NL 6)

The above suggests that a shift from looking at subjective signs and symptoms of individuals to objective tests with most, if not all, pregnant women in the community, accelerates the proliferation of these technologies into previously untouched social spaces. While genetics is being increasingly shaped by strategies of surveillance medicine, it is a belief among some physicians that the health of a community will only be ensured if the whole of the population comes within the range of genetic surveillance. Within this paradigm, there could arise a danger that genetic prediction will become linked, if not equated, with disease prevention. Perhaps this link is already being established, as one policy maker suggests:

> People are now interested to ask questions about prenatal diseases. They are interested in knowing what kind of child they will have ... because abortion is possible ... If there is a psychological or any other problem illness ... abortion is possible ... People know that there is a system ... a way to determine the health of the child ...
>
> (G 10)

The scientific scope for the development of genetic technologies may appear endless as techniques, the range of knowledge and knowledge interests, and work organisation within surveillance medicine continue to grow. Genetic technologies embody complex social relationships and experts and patients involved in these relationships are a structuring piece of surveillance medicine, the cultural configuration in which genetics is currently being reproduced.

Genetic risk identity as novel

Very often the relationships between genetic experts and their patients originate in the quest for risk assessments in order to avoid potential illness. Patients may ask: 'Am I at risk of getting this disease?' or pregnant women may ask: 'Is my 'baby' at risk of being born sick or disabled?' In this context, Armstrong (1995: 403) contends that the importance of surveillance medicine lies in the manner in which the machinery of scrutiny deployed throughout populations to monitor 'precarious normalities' delineate a 'new temporalised risk identity'. Simply, new risk identities are constructed through the application of biomedical knowledge and practices, which mark out an individual's inherent eventuality for future illness potential. I would contend that the delineation *par excellence* of this new risk identity becomes most visible in genetics which most definitely enhances 'the culture of risk' (Kerr and Cunningham-Burley 2000). Here, genetic risk identity becomes a novelty in biomedicine. Indeed, the construction of a genetic risk identity is qualitatively different than other types of risk identities in biomedicine. This construction is novel because only with the discovery of genes can risk determinations based on genetic make up or DNA based material in bodies provide the statistical probability and in some instances, certainty, that an abnormality or disease will occur. A genetic risk identity is all about the heritable, hidden trait embedded in a carrier's body, the hazardous corporeal material that makes you – you and the embodied, innate danger, lurking in your inner, living spaces. Through the powerful scopic drive in medicine, genetic risks can be seen to be embodied in a tangible individual with identifiable 'hazardous' genetic material.

With regards reproductive genetics, that permeable boundaries exist between precarious normalities and future diseases and that this threat exists in the womb because these boundaries can be breached are relatively novel notions for pregnant women. Pregnant women's self-identities and experiences of their sometimes anxious, gendered bodies can also be shaped by expert's risk calculations. Here, genetic risk takes on a contingent character in reproduction and is envisaged as embodied in women and jeopardising one's future womanhood as a good reproducer. In their construction of risks, experts bring what is interior and central to the pregnant body to the outside for medical perusal, marking specific 'defective' bodies as distinctive. When 'faulty' genes or problematic foetuses are discovered, genetic risk becomes a deeply gendered experience, a painful reality to be confronted as a private choice in considering what to do about one's body and one's foetus. In this context, Lippman (1999) contends that pregnant women's choices are framed as an individual consumer's self-expression rather than a medical necessity. This revelation of our interiors and constructions of embodied risk are technological practices leading to cultural and individual reflexivity (Lundin 1997). Indeed, the monitoring of risk is a key aspect of reflexivity (Giddens 1991: 114) in society generally. Specifically, in reproductive genetics, genetic risk monitoring offers an illustration of routine reflexivity and the embedding of risk assessments in interactions between experts and pregnant women.

One expert, a clinical geneticist, describes this process of interaction and how important prenatal procedures are in order to inform 'every patient' of their 'reproductive risks'.

> Prenatal diagnosis has been used as an instrument in the health care case by case between the doctor and the patient. So it's good medical practice to inform every patient of their reproductive risks and the options ... for prenatal diagnosis by performing highly reliable tests [for] at risk pregnancies.
>
> (NL 4)

Nevertheless, when talking specifically about prenatal screening he has reservations and reveals that he is not happy about some of the consequences of these prenatal procedures. For example, he is worried about the expense, organisational efforts and creation of anxiety:

Risk information is very complex ... [In] a centralised lab, they ... have a computer ... printing out relative risk values ... You'll have to start with the number of exceptions and influence of maternal weight, race, diabetes and exceptional levels of either very high or very low hormonal levels and so on. [This] makes these tests ... extremely ... complicated ... [With] tremendous expense and organisational efforts, you are creating a lot of anxiety in up to 2 per cent to 5 per cent of pregnancies, necessitating ultrasound, follow-up testing ... amniocentesis and so on.

(NL 4)

He continues by mentioning one of his colleagues who is very opposed to prenatal screening (i.e. maternal serum screening). He says that his colleague's fear is that the public would have a negative view of prenatal screening. In a real sense, the routine reflexivity of prenatal screening, while being a beneficial risk calculator for experts, becomes problematic on a cultural level when it is viewed as consistently undependable:

He has always been very opposed to less reliable tests [sic] in pregnancies, fearing that less reliable test methods for at risk pregnancies would become viewed by the public in general as messy and unreliable.

(NL 4)

In this context, monitoring of genetic risks through prenatal screening does not provide ontological security for pregnant women. Rather, these risk determinations are based on mathematical possibilities, not medical certainty. A clinical geneticist, knowledgeable of the unreliability of these prenatal screenings suggests another major problem with serum screening – its incomprehensibility for those interpreting the results as well as pregnant women. The following excerpt reveals that while prenatal screening produces supposedly reliable computations of the probabilities of carrying a Down's syndrome foetus, the routine reflexivity of the speaker challenges this medical view. For her, risk calculations and risk results are not only complicated but also confusing:

Down's screening is a very complicated screening that practically no one really understands it ... You screen to find a group that has a higher risk ... If I am pregnant and go to a screening, they can't tell me whether I will have a Down's child or not. But, they will tell me something about my risk of having Down's child ... According to that, I decide whether to take the test or not. It is ... very complicated because even if I get very good results ... I still may carry a Down's child. And the other way around. Even if I get the worst results, it is rather likely that the child is healthy.

(F 5)

Referring to maternal serum screening, the same expert felt that it was very unfair to give women 'false information' during pregnancy. She continues:

A problem was that [maternal serum screening] gives false infor- mation. Then somehow it's very unpsychological [sic] and cruel during the pregnancy to start to inform ... you of everything ...

(F 5)

On the other hand, a gynaecologist 'defended' prenatal screening. He was aware that some of his colleagues were not favourable towards it, when he asked, 'What is the big thing?', which I interpreted to mean, 'What is the big deal with maternal serum screening?' This expert's ideas on the monitoring of risks illustrate an attitude that, in fact, does not minimise the difficulty of making decisions based on risk calculations. Yet, in a subtle way, he contradicts himself because he says he does not believe prenatal screening is a 'big thing'.

We defend ... that what we do is select a subgroup ... at risk for congenital abnormalities. [This is] based on their age [and] is a type of screening ... Whether you tell me your age or whether I tell you ... [You] give me a bit of blood. Basically this is the procedure. Knowing your age is ... [determining] a preliminary risk ... Doing a blood test in the 16th week ... getting a result after a couple of days – then having to decide amniocentesis or not based on your individual risk calculation is not an easy thing to do. So if you ask me what is the big thing? Why is it a big thing?

(NL 2)

The above expert also demonstrates that while risks may be measured objectively, the revelation of good results and little or 'no' risks may lead a pregnant woman to a type of embodied reflexivity in which her experience of reproduction is traversed by a sense of personalised risk almost living or embedded deep inside her own body. When total certainty from technological practices is not possible for a pregnant woman, rationalising her genetic risks may mean taking a test as a further step towards bodily certainty. The following expert reflects on the sort of bodily certainty that pregnant women want:

> You have this individual. Let's say [there's] individual chance. It's an objective thing. It's a chance. A risk is a subjective thing. It is something in your mind – the kind of risks you run. There's a woman. I congratulate her [about] the result of her test ... Her risk is only 1 in 1000 and she says '1 in 1000, I know what to do, I want to have amniocentesis. Because that 1 is me'.
>
> (NL 2)

Nevertheless, pregnancy is never risk free – with or without technologies. Whilst this may be true, institutional pressures protect the established status of any expert and his/her judgements which become used to control the discourse about risk (Price 1999: 51). In this context, one expert speaking about prenatal screening and diagnosis suggests that in medicine there are no tests that guarantee a perfect baby – no guarantees that your pregnancy will be risk free. She implies that a woman should not get pregnant if she cannot deal with risk:

> When ... you find that you give birth to a child who is going to die or who is going to be disabled ... your world just crumbles ... It is that sort of person who wants the guarantee ... It's so difficult to get the person to understand that there are no such tests and that if you don't want to take a risk, you should not have a baby.
>
> (Clinical geneticist F 17)

What becomes clear in analysing the notion of risk *vis-à-vis* prenatal screening specifically is that risk calculations are not only uncertain but also change. The following expert, an epidemiologist, suggests that changes in risk calculations are based on the capacity of a particular service to screen effectively and organise support services (i.e. counselling). In other words, the external services should determine the numbers of 'risky' pregnancies who will receive further help from support services. For him, screening is 'soliciting trouble'. However, his moving of the risk calculation 'cut off' figure implies not only do his constructions of risky pregnancies alter but also that he may be unwittingly 'soliciting trouble' by moving the risk calculation 'cut off':

It's the wrong perception of screening to think that [it] is there and
... then [it] creates a consequence. You decide what consequence
you could cope with and then you adjust the screening and that's
so important ... Screening is actually soliciting trouble and you
don't want to solicit more trouble than you can cope with ... In
screening you can control the amnio rate ... You just move the
cut-off, so that you can accommodate the counselling and the
amnios ... All you have to do [is say], 'I can't cope with 5 per cent
of amnios because I haven't got the staff to do it or laboratory'.
So you move it out to 2.5 per cent and then the detection rate ...
won't go down that much.

(E 6)

In a similar way, another expert, a clinical geneticist, also implies
that her risk calculations change when she speaks about 'adjusting the
risks':

In serum screening, we have a limit of 1 in 350 at birth for Down's
syndrome ... There is a difference if you say the risk [is] now in the
middle of the pregnancy or at birth. Well ... if you have a Trisomy
21 pregnancy then there is a risk of miscarriage during the end of
it. So 1 in 250 now equals 1 in 350 at birth. So these risk figures
might differ in numbers but they might be the same in the end ...
   You always adjust your own risks. We try not to have too many
screening positives. Five per cent of all ages will be considered ...
the accepted goal ... Then you estimate that about 60 per cent of
Down's syndrome will be found. So it's an adjustable limit.

(F 16)

As we saw in Chapter 1, besides serum screening, pregnant women
and their partners can also be screened through the use of molecular
genetic tests. In talking about these sorts of developments in genomics,
one clinical geneticist saw this as a 'revolution'. She says:

A lot of this research is really beneficial to these families and you
can really say that it is a revolution. We have a lot more to offer to
these families nowadays than 5 or 10 years ago, but still I do not
think that everything is going to be just good. There is also something that is going to be risky and testing for susceptibility is a
most difficult thing to cope with.

(F 4)

In this context, experts look sometimes to whole populations as a way of identifying genetic risks and they organise objective tests in communities. Here, one policy maker wanted to see the use of genetic tests for the detection of rare diseases. When focusing on identification of rare genetic diseases in the population, he used the metaphor of 'rare postage stamps', suggesting that these diseases should be seen as out of the ordinary:

> There are special genetic diseases, but they are rarities. They are special problems handled by the medics in their very clearly defined settings ... It is rather like special stamps that you collect and must be seen as such as special ... There are ... good examples where screening ... families where one knows that there is a risk ... we can advise the parents ... and counsel them ...
>
> (F 13)

Another expert, a paediatrician, was concerned with population screening for thalassaemia. His work focused on detecting young couples at risk.

> [We] try to screen all couples ... This is actually population screening ... we are ... detect[ing] the couple at risk ... We try to offer to [them] the option of only having the child by doing antenatal diagnosis. So if there is an affected child, they have to go for termination of pregnancy.
>
> (G 6)

He goes on to explain how one has to be 'careful with counting' and analysing mutations:

> You have to be more familiar with the techniques – the different variation of models because there are several types of mutations, so you have to be careful – careful with counting ... In these centres, the diagnosis is quite efficient ... The DNA technique is much more accurate ...
>
> (G 6)

While, as noted above, experts look to entire populace as a way of identifying genetic risk, routine reflexivity must also involve what one expert called 'risk setting' (i.e. generating risk awareness about genetic disease) for the population. 'Risk setting' is important because monitoring of risks may cause fateful moments (Giddens 1991) or moments highly consequential for a population's destiny. This expert felt that public discussion would be helpful in 'risk setting'. She says:

> We have 1 in 10 people ... [who] will be carriers [of rare diseases] ... In fact, you will get people scared ... I think we should have a sort of informed discussion about this issue and ... If information is ... served to the public, it should be served in a ... comprehensible way ... in a true way, so that it gives ... the true risk setting to people.
>
> (Clinical geneticist F 17)

In terms of risk setting for the public, one clinical geneticist envisaged the monitoring of genetic risks as an activity soon to be embedded in daily medical practice. However, she believed that this sort of routine reflexivity which focused on genetic risk identification would be extended to others in the medical profession and not only the specialists. The implication is that more members of the medical profession will engage in genetic counselling of their patients and communicating risks:

> If we foresee the future ... you can identify risk individuals for hypertension – that it is possible that somebody else [other] than the specialist of paediatrics or internal medicine gives that information ... I think this is a big challenge for the next 10 years ... for the medical genetics profession ... The whole counselling concept should be changed to be more informative and more informing the general public and the test users.
>
> (F 3)

Another expert, an ethicist, was extremely cautious about the new developments in genetics. She focused on the monitoring of genetics diseases and the use of sex selection when an at-risk pregnancy was identified. She believed that sex selection when one was known to have a genetic risk has profound ethical implications. She says:

> Sex selection is probably unethical except where there is risk ... We will have to look very carefully at [these issues] but we ... certainly would look at the area of ethics ... So much discussion is going on here ...
>
> (E 7)

The above excerpts demonstrate how genetics risk identities can be viewed as a novelty, when experts' technological demarcations

construct these identities for pregnant women as well as the population. In the area of reproductive genetics, pregnant bodies, touched by
prenatal technologies construct risk awarenesses, regardless of whether
or not their encounters with these technologies reveal diseases or
abnormalities. Experts' monitoring of genetic risks reveals routine
reflexivity in prenatal and public spaces. Encountering a risk culture,
pregnant women may experience an embodied reflexivity and desire as
much technology as they can have in pregnancy for a sense of bodily
and reproductive certainty.

Reclaiming the genetics agenda: containing diversity and shaping risky bodies

In this and the previous two chapters, we have looked at the complexities of genetics in terms of the various procedures and technologies
used, its knowledge base and knowledge interests and its work organisation in surveillance medicine, constructing genetic risk identity as
novelty. Here, I would suggest that social scientists who attempt to
analyse the genetics and specifically, reproductive genetics discourse
must be aware of the power that surveillance medicine and expert's
constructions of risk have in determining the genetics agenda in
society.

In this context and stemming from her concern about geneticisation, Abby Lippman (1992) asks what if we changed genetic
metaphors and for example, see a gene map as no more than an
organogram (i.e. the grand bureaucratic design of an organisation –
fixed and orderly). She asks further: 'Do we really want to invest major
human or economic resources in the development of an organogram?'
(Lippman 1992: 1474). Her main aim in asking these questions is to
call for a reclaiming of the genetic agenda by a thorough exposure of
the genetic colonisation of health and illness in and through the narra-
tive of contemporary genetics.

In the process of reclaiming the genetics agenda, we as social scientists need to contextualise the human body as a politically inscribed
entity, its biology and pedigree shaped by histories/herstories and practices of containment and control as well as difference. In our analysis,
we need to place the body at the core of political struggles (Turner
1996: 67). Bodies need to be seen as sites where the knowledge of
genes, foetuses, reproductive functions and the universalising system of
surveillance medicine with genetic constructions of risk converge and
not as gender neutral, non-determinate systems. One difficulty is that
this work needs to be done with the explicit intention of demon-

strating how notions of 'genes in bodies'; the paradigm about healthy and diseased genes; and ideas about appropriate kinds and levels of reproductive performances are culturally dependent 'embodied processes'. Thus, this work is about the need for a resurrection of the body in our work and the breathing of 'epistemological' life back into our neglected frames. Our work is about the affirmation of corporeality, making the distinct claim that the body exists very centrally in the genetics discourse. To make this claim is to begin to understand why specific genetic metaphors have been needed not only as rhetorical tools but also active agents in forming a genetic 'moral order', which is dispersed in a multiplicity of ways with far reaching effects.

4 Shaping pregnant bodies

Distorting metaphors, reproductive asceticism and genetic capital

> Patriarchal childbirth – childbirth as penance and as medical emergency – and its sequel, institutionalized motherhood, is alienated labor, exploited labor, keyed to an 'efficiency' and a profit system having little to do with the needs of mothers and children, carried on in physical and mental circumstances over which the woman in labour has little or no control.
>
> Adrienne Rich, 'The Theft of Childbirth', p. 163

Introduction

In many countries today, medical professionals maintain that most, if not all, women who are pregnant should engage with the techniques of reproductive genetics. At the same time, research indicates that men and women in the general population believe that these sorts of techniques can be useful in eliminating a variety of serious genetic diseases (Marteau 1995; Marteau and Drake 1995; Marteau *et al.* 1995; Newberger 2000). As these social trends in the public's perception of genetics materialise, important moral questions are being raised (Rhodes 1998; Cunningham 2000; Chadwick 2000; Spallone *et al.* 2000). For example: What diseases will be perceived as life threatening or grave? Will drug addiction, alcoholism, homosexuality, schizophrenia, severe depression and so on appear on lists of serious diseases? Will poverty and unemployment somehow be included in this disciplinary process? Will hierarchies of social diseases and bodies be created? If so, who will decide on these hierarchies? Could these hierarchies be used as a basis for social bias or discrimination? Is there an already invisible hierarchy that has developed from the use of prenatal technologies? Most importantly, how do all of these prenatal technologies affect the pregnant body, the major target of prenatal technologies?

In an attempt to provide an answer to the last question, this chapter begins to develop a feminist, embodied approach to reproductive genetics. A feminist, embodied approach takes seriously women's corporeal experiences, recognises that gender is embodied and that material changes in women's bodies, such as pregnancy, may have implications for gender identity (Bailey 2001: 111). Reproductive genetics is not just about the surveillance of pregnant bodies within the institution of reproduction. It impels women to perform 'correctly' their pregnant bodies in the normative institution of reproduction. Pregnant bodies are directed by physicians to play out their reproductive roles in biomedically approved ways, as these bodies are pushed into the service of technology. Especially older pregnant bodies are drawn into a prescriptive discourse, outlining the correct paths to follow for a successful pregnancy (Beaulieu and Lippman 1995). The potentially problematic nature of pregnancy, the 'managing risk repertoire' of pregnant bodies and the potential for foetal pathology or genetic disease are used as justifications advocating prenatal testing and medical intervention (Marshall and Woollett 2000).

In her work on the need for embodying theory, Kathy Davis (1997: 14–15) says that feminist theory on the body provides an essential corrective to the masculinist character of much of the new body theory. This is because feminist theories start off by analysing difference, domination and subversion as a way of looking at the conditions and experiences of embodiment in society. Asserting that bodies are not generic, she elaborates two problems that one confronts in attempting to develop embodied theories on the body. The first problem relates to grounding theories of the body in the concrete embodied experiences and practices of individual women and the second is the need for 'reflexivity of theorising the body'.

I attempt to deal with both of these problems in this book. Firstly, my theories around making sense of reproductive genetics are grounded in the assumption that pregnancy is a deeply embodied experience, albeit a gendered experience. Secondly, reflexivity of theorising the body comes through when pregnant women's subjectivities constituted by their embodied experiences and different reproductive choices are seen to be constructed by a variety of constraining expert discourses. A reflexive theorising of the pregnant body includes a theory on the lived body – the female body shaped by gender, cultural norms and expectations as well as lived experiences.

In this context, Rosi Braidotti (1994) offers a feminist focus on the embodied subject, specifically within an analysis of the medicalisation of the female reproductive body. She uses Foucault's idea of embodiment, what she calls bodily materiality, and contends that this notion, the materialism of the flesh, defines the embodied subject as 'the material concrete effect, as one of the terms in a process of which knowledge and power are the main poles' (Braidotti 1994: 57). For her, contemporary biomedical sciences have acquired the right as well as the essential knowledge and expertise to act on the very structure of living matter. In this process, they extend their control over the depths of the maternal body. This has caused a multiplicity of problems on an ontological level for women, given that making babies is a major concern for an ageing, post industrial Western world and, as Braidotti says, 'wilful reproduction is in' (Braidotti 1994: 51). That the embodied subject, especially the maternal body, has become fragmented or 'broken', as it has developed into the target of biomedical discourses and subject to intense scrutiny by its powerful scopic drive, is a major problem which Braidotti exposes.

Indeed, Davis and Braidotti, as feminist embodiment theorists, expose the need to contextualise and understand the conditions and experiences of embodiment *vis-à-vis* living bodies. In this context and with regards reproductive genetics, we should envisage the procreative body and more specifically, the reproductive body as the end product of a whole system of cultural relations, power and knowledge. Mirroring the body in technology, the pregnant body in reproductive genetics is 'speaking in new ways' (Andrews and Nelkin 1998) as serum, blood and genetic materials are being extracted from wombs and foetuses to be used in statistical calculations of risk or visualised in DNA sequence variations. Here, a discovery of the sorts of difficulties and tensions which the pregnant body in reproductive genetics evokes is in order.

As well as offering a feminist embodied approach to reproductive genetics, I want to generate in this and the following chapter an awareness of some of the disciplinary practices in which pregnant bodies, seeking 'good foetal bodies' and eventually healthy, if not 'perfect, babies' are involved. I want to highlight key feminist and sociological issues that become visible when the body, specifically the pregnant body, becomes a theoretical site. In order to position the body as a central feature in feminist and sociological analyses of reproductive genetics, the chapter includes three main discussions. Firstly, I will look at how genetic metaphors, as powerful metaphors, not only shape distorted, unnatural conceptions of the human body but also deny the fact that bodies are inscribed by categories of difference, such as gender and disability. Secondly, I will look at the mechanistic view of the body in reproductive genetics and its effects. Thirdly, I will examine how reproductive limits are practised on pregnant bodies through a feminised regime of reproductive asceticism.

The central assumption, running throughout this chapter, is that the science of genetics is a part of disciplinary process, offering a limited view of the human body. This disciplinary process tends to conceal that what may appear as 'flawed genes' is in fact a result of a body's interactions not only with the environment but also gendered social practices valorised by difference as well as rigid definitions of health and illness.

Genetic metaphors: shaping distortions of the body

The circulation of symbols, images and myths about genes 'in bodies' permeate the fabric of contemporary society. Used regularly, metaphors signal the proliferation of genetic knowledge and technologies into our everyday life. Genes bring to mind potent images and are types of universal signifiers. Nelkin and Lindee (1995a: 6) contend that there are three related themes that underlie the metaphors geneticists and other biologists use to describe work on the human genome. These include the characterisation of the gene as the essence of identity; a promise that genetic research will enhance prediction of human behaviour and health; and an image of the genome as a text that will define the natural order. They envisage the gene as a cultural icon, exalted above everything else in human nature; a type of omnipotent signifier. In related contexts, Hubbard and Wald (1993a) attempt to explode the gene myth and suggest that the search for the human genome can be likened to the quest for the Holy Grail, a quest for the sacred object of Western civilisation which ended up as a prolonged pursuit carried out mainly by privileged men (knights) during the Middle Ages. Other scholars (Rosner and Johnson 1995: 104), familiar with these kinds of genetic metaphors, suggest that, whether living (i.e. stories) or dead (i.e. inanimate objects), these metaphors 'invariably cast the scientist as the one who dominates and exploits the Other'.

While science has consistently presented itself metaphorically, representations that are seemingly objective are, at the same time, forays into the ways in which scientific thinking proceeds along both traditional and experimental networks (Anker 2000). The science of genetics is no different. Rather than belabouring the point that genes bring to mind powerful images or are types of universal signifiers, I would like to take a different tack and proceed to a discussion of how genetic metaphors, as authoritative metaphors, shape somewhat warped or contrived conceptions of the human body as well as deny the fact that bodies are inscribed by categories of difference and subject to strategies of attention (i.e. power) which surround them

(Armstrong 1987: 66). In this context, three metaphors will be discussed: genetic landscape, genetic mapping and genetic blueprint.

Genetic landscapes, genetic mapping and genetic blueprint: spatialising genes

In their anthropological analysis of the genetic history of the Iberian Peninsula, Calafell and Bertranpetit (1993: 736) refer to various sizes of genetic landscapes: 'the world, Europe, Italy, the Iberian Peninsula'. Here, the image is one of places where groups or individuals with specific genetic pedigrees migrate, inter-mingle, copulate, confront barriers, slip through boundaries, drift and populate. The word, 'landscape' itself depicts an actual segment of inland scenery: a place, or a piece of art reproducing this inland scenery. Landscape can be about one's current line of vision but it can also be about one's past history. There is always something to be found below the surface of a landscape. Through landscapes, we may find out how others have travelled, lived and passed through a particular locality. In this sense, a landscape becomes a useful tool for reconstructing our past.

Another metaphor, gene mapping, conjures up images of a guide showing us the way. One author (Culliton 1990) has suggested that what is being mapped is really unknown territory, 'the terra incognita human corporis'. Generically, mapping or more precisely, a map, refers to 'representation of earth's surface showing physical features' (*Concise Oxford Dictionary* 1995). Nowadays, with constant border changes in many parts of the world, maps become representations of the earth's surface showing not only physical but also political features, reflective of successful attempts of powerful states to transgress boundaries. In this light, gene mapping not only suggests location but also control and ownership.

In other contexts, we are told that genes, or more specifically the DNA molecule, constitutes a genetic blueprint. A blueprint is literally 'a blue photographic print representing final stage of engineering or detailed plans of work to be done' (*Concise Oxford Dictionary* 1995). The key here is that blueprint brings to mind the notion of work to be done, similar to a recipe (Condit and Condit 2001). At their most valuable stage, blueprints are 'in progress' plans concerning localities. Furthermore, blueprints suggest that genetic developments are somehow due to the purpose or design that is served by them (i.e. genes are sets of 'final causes') and thus serve a teleological purpose.

While these metaphors hint at the scientific need to reconstruct the past; to control and own our present; and to develop our future, what all of these metaphors have in common is that they cast the scientist as the one who dominates and exploits the other (Rosner and Johnson 1995), conflicting with feminist values. In this context, Kathy Davis (1997), referring to Evelyn Fox Keller's 1985 book, *Reflections on Gender and Science*, contends that in a real sense, the female body becomes itself a metaphor for all that needed to be tamed and controlled by the dis-embodied male scientist. While metaphors serve to simply complex information and facilitate the communication process (Petersen 2001), the above genetics metaphors suggest the idea that genes can be precisely spatialised; captured in a gaze, charted, or measured. Simply, this sense of spatialisation is about what can be termed, genomic locality. On the one hand, every scientist knows genes are located in bodies. On the other hand, in order to hallow genes and empower all sorts of new technologies, scientists need metaphors, protecting the primacy of the gene – not the body.

As Gottweis (1997: 57) exclaims: 'We are our genes. Eureka!' In this logic, scientists and experts uphold the view that genes – not bodies – determine human social behaviour. They believe that genes, active in various combinations rather than bodies, inscribed by culture, are fundamental in explaining social performance. As biology and genetics are pushed into the foreground of social consciousness, culture takes a back seat, if any at all. In effect, this process obliterates the fact that bodies are marked by categories of difference (gender, race, disability, class, etc.). It establishes the body as 'but an epiphenomenon of its genes' (Gilbert 1997: 40). In the sphere of reproductive genetics, the bodies of pregnant women and the disabled fall outside of the paradigm of health or well being, as the former bodies are being invaded while the latter are being eliminated. Medical technology drives this process, whether living pregnant bodies are being occupied or potential disabled bodies done away with.

Here, it is important to remember that the object of a 'seeing science', the human body, is already a cultural object invested with meaning (Newman 1996: 4). Furthermore, scientific conceptions create notions of genes endowed with agency, autonomy and causal primacy (Fox Keller 1995). At the same time that they are establishing the primacy of genes, scientists and experts create and need metaphors that project genes outside of the body and project the body outside of one's perceptual field. Inanimate images, such as landscapes, blueprints, maps, are made to appear or at least function as animated objects, while the living corporeal body, the site of genomic activity, is denied agency. This material body becomes non-existent. Indeed, metaphors serve to exclude the body from the genetic gaze. Through the powerful scopic drive of science in the form of reproductive genetics, the body becomes a non-entity, a non-thing – a nothing. At least, if a body is dead, it exists and can be seen. But, if the corporeal body is excluded from the perceptual field of science, it becomes as it is perceived to be (Armstrong 1987: 66) and exists in the discerning mind not in observable reality.

The power of the above and other recurrent metaphors found in the discourse of genetics acts like pervasive mantras re-working ever so methodically not only the scopic field in which scientists perceive, experience and legitimate genes, but also the socio-spatial context in which bodies, both living and dead and marked by difference, can be excluded, if not eliminated from genetic consciousness.

Mobilising concepts and dispersing the genetic moral order

Besides metaphors, powerful genetic concepts, such as risk, affected offspring, viability, defective genes, carrier status and so on are mobilised by experts to mark 'harmony' and 'disharmony' in the progession of a woman's reproductive process and ultimately, between herself as a pregnant woman and her foetus. These concepts help circulate a genetic moral order. In this emergent moral order, the development of new technologies affect perceptions of our ability to control health, illness and reproduction as well as produce new fantasies on how these technologies can be used and abused (Van Dijck 1998). Within reproductive genetics, gender relations are embodied in and reinforced by technologies that are designed especially for pregnant bodies. In this context, Faulkner (2001) argues that an embodied form of gendering is going on with these technologies, as they have material consequences for gender relations. These technologies encourage a tendency for women to become disembodied – to reduce them to their wombs – because these technologies lessen physicians' reliance on women's knowledge and experience (Faulkner 2001: 84). Located firmly within the politics of reproduction, prenatal technologies may expand pregnant women's options as they want healthy babies, but clinicians who use these technologies tend to promise more than they can deliver (Casper 1998: 81).

The fact that a pregnant woman is a subject situated within a labouring body with her own point of view (Sbisa 1996) tends to be minimised by clinicians within the institution of reproduction. In this context, Susan Bordo (1993a: 93) contends that not only are women's reproductive rights being contested but also their status as subjects within cultural arrangements which, for better or worse, the protection of the real subject (i.e. foetus) remains a central value. Looking over the reasons why the contemporary language of reproductive genetics has such persuasive social power, one can see important, sometimes hidden, social processes and disciplinary practices being played out.

The body as machine: dangerous genes and foetal containers

At heart, the language of reproductive genetics relies on and privileges an individualistic, mechanistic view of a gendered body with the effect that the full significance of reproductive processes is lost. Of course, this view, modelled on the workings of an inanimate object (i.e. a machine) is not new in medicine, as discussed earlier. Peter Conrad (1999) argues that the notion of the body as a machine emerges from germ theory that is the basis of the clinical medical model, underlying biomedical and public thinking about disease. Within this paradigm, the body is treated as a machine and it is the doctor, the mechanic, who fixes it (Martin 1992). Conceptions formulated within 'the body as machine' viewpoint help to maintain gender bias rather than gender neutrality (Mahowald 1994). Indeed, the science of genetics is a gendered social practice and as all other gendered social practices (Lorber 1997: 3), it is capable of transforming the body. We are able to see this sort of gendering happening when masculinist science re-conceptualises reproduction as a technological rather than natural process. In this way, feminine ontology is intrinsically compromised (Shildrick 1997). Most importantly, the notion of reproduction as a valuable, material site of embodied experience for women vanishes.

On another level, Leder (1992) suggests that side by side the thrust of mechanistic and technological interventions in medicine a schizophrenic shift is taking place as physicians see the body as a machine and at the same time, as belonging to a living, breathing person. Leder contends that in order to attend to the living body, thoughtful scholars must refuse to grant 'this mechanical wisdom the status of ruling paradigm' (Leder 1992: 31). Nevertheless, within the context of reproductive genetics, not to grant this mechanical wisdom a privileged place is fraught with problems, as one expert points out:

> We all believe in genes, biology, the body and [the] social condi-
> tion. But the problem still remains that we have a scientific
> mindset which means that we privilege the kind of information we
> get from objects and events over these more complicated relation-
> ships and processes ... So once you have genetic screening

technology your problems are solvable by genetic screening technology and then you don't think creatively anymore ...

(Researcher E 1)

Treating the body as a machine is consistent with the belief that
genes can be manipulated, undergo therapy or be altered. We have the
idea that bad or defective genes can be weeded out from the good ones
and replaced. In the end, problems arise as attempts are made to make
hard and fast distinctions between good and bad genes, as one can
with, say, a good and a bad car. A clinical geneticist discussed that one
effect of making this distinction between good and bad genes was to
ask oneself: 'Where does one draw the line?'

We are wanting to go [in] ... that direction that we are selecting for
the good genes and selecting against the bad genes. Where is the
limit between good and bad genes? That is the question. Because
there are examples where this limit is really hard to draw.

(F 4)

Within this body as a machine paradigm, the body is also viewed as
a container of genes, a carrier of genetic material. This idea perpetu-
ates the belief that genetic disease is not only something terrible that
someone has but something someone is (Steinberg 1997: 117). If genes
are diseased, someone is diseased and this someone's body can be
perceived as carrying dangerous material. In this context, one expert
says:

I think that very often people get afraid. They think they are
carrying something that is very dangerous and so on, and at the
same time, people ... don't understand that we all have genes which
are more or less diseased or which might get a disease or so on.

(Lawyer F 15)

In the reproductive context, the pregnant woman is viewed as the
receptacle of her own and her future offspring's genes, a foetal
container (Purdy 1996: 88), a view endorsed by the medical profession
(Ward 1994). While bits of DNA combined to make genes represent
updated versions of seeds, the reproductive processes and experiences
of maternity and maternal ties hold little if any significance in
comparison to paternal, patriarchal ties (Katz Rothman 1995). This is
not surprising given that knowledge of paternity has served as a
primary measure of progress towards civilisation (Franklin 1997: 25).

Thus, paternal – not maternal – ties are imputed to be based on genetics. On the one hand, blood ties are meant to imply genetic ties, connections made by seeds, semen or men. On the other hand, maternal ties, based on wombs and the growing of already born children play a less important role in genetic progress.

The pregnant body, reproductive asceticism and genetic capital

The language of genetics sets reproductive limits both upon the inner body and the outer body in our modern consumerist culture with the result that women's, more than men's, bodies are restrained. In short, the science of genetics becomes an ideal way of bringing together what Turner (1992: 58–9) has referred to as external problems of representation (i.e. commodification) and regulation and the interior ones of restraint (i.e. control of desire, passion and need) and reproduction. What does this mean? Simply, through reproductive genetics the pregnant body will experience self-imposed restraint through a type of reproductive asceticism. While 'pre-pregnancy' tends to involve self-disciplining oneself and one's female body, pregnancy itself is regulated through the ways in which normal female bodies 'fit' to reproduce are construed (Marshall and Woollett 2000: 363).

The distinct, free, embodied, 'natural' process that should shape pregnancy or the liberatory performance of female procreation, is subjugated through the routines of a sort of 'technological' reproduction shaped by genetics. Both physical and technologically induced changes in pregnancy are embodied experiences that can, interestingly enough, be seen as both resources shaping gender identity (i.e. future motherhood) and forms of social and biomedical control, reducing women to their biology (Bailey 2001). The norm of a free, embodied pregnancy has been replaced through the increased medicalisation of reproduction. The discipline of reproductive genetics aids in this medicalisation process through the circulation of its limiting but powerful routines, values and invasive practices. Pregnant bodies are viewed ever more as immaterial, while at the same time, these pregnant bodies and their physicians are brought into a system of normative surveillance through the reign of technologies (Balsamo 1999).

While prenatal technologies can be seen as a socio-technological system (Cowan 1994), in this system, prenatal genetic testing may become a need (Beaulieu and Lippman 1995), a lifestyle or even an addiction (Lippman 1994) in our consumer cultures. Here, the image of the foetal body comes to signify endangered childhood in need of parental protection (Taylor 1993) as well as a commodity or product subject to quality control (Rapp 1999). In her discussions of female consumption, Bordo (1993b: 196) contends that consumer capitalism depends on the continual production of novelty and fresh images to stimulate desire. In this sense, the continual development and deployment of technologies within reproductive genetics incites pregnant women's yearning for perfect offspring. As disciplinary and consumerist practices, these technologies become embedded in reproduction, stimulating pregnant bodies to perform well. A continual question arises for the pregnant woman. 'Is my pregnant body and are my genes "good enough" to deliver a fit baby?'

On the level of morality, Taylor (2000) maintains that consumption itself to a considerable extent constitutes the experience of pregnancy and is invested with a new level of moral significance. For her, consumption is 'cast as an act of maternal love, an expression of a woman's strength of character and powers of self-discipline, even as consumption is seen to literally create the foetal body' (Taylor 2000: 403). Pregnant bodies are consuming bodies; pregnant women 'consume' prenatal technologies for successful pregnancies as well as for the well-being of their foetuses. Within the disciplinary matrix of reproductive genetics, both maternal love and reproductive asceticism become one. Pregnant bodies are constructed as those who should engage in 'exemplary' self-disciplinary and gendered consumerist practices in the genetics moral order.

As prenatal technologies become routinised, pregnant women begin to accept these technologies under the rubric of older non-controversial medical practices (Press and Browner 1997). On the one hand, some pregnant women may have almost total trust in their attending physicians (Kabakian-Khasholian *et al.* 2000) and thus engage enthusiastically with prenatal technologies. On the other hand, others' experience of their own expectant body (Rapp 1998) suggest that theirs is a measured trust in a system of normative surveillance.

Whether pregnant women experience total trust or measured trust, genetic uncertainty often becomes played out on their bodies. Thus, in the world of reproductive genetics, pregnant women come to embody uncertainty. In this context, uncertainty is a paramount source of fear in society: it is the lack of control over so many unknowns within the life equation of an individual (Bauman 1987: 10). Shaped by the new genetics, uncertainty is a health as well as an ethical issue (McCaughey 1993), given that more defective genes than ever before are being identified. In the social mix between genes, gendered bodies, personhood, health and disease, a cultural craving for intellectual security emerges.

Identity itself becomes a genesis problem (Heyd 1992: 160). Here, the need for certainty and security is linked to various anxieties situated in large social frames, interpreting difference *vis-à-vis* 'the genetic master code' as a threat (Hartouni 1997: 119). In turn, anxieties are mediators of micro–macro relations in the sphere of genetics and construct 'emotion' narratives (Williams and Bendelow 1998: 47) surrounding this discourse. In reproductive genetics, emotion narratives are stories of dynamic interactions mixed with organic, genetic, psychosocial and gendered processes between pregnant bodies, their world of social relations, their environment and the institution of reproduction. As these narratives unfold, they become raw data for experts: 'the stuff' from which professional concerns surrounding the social relations of reproductive genetics are being worked out as well as tales of insecurity, uncertainty and measured trust.

One clinical geneticist explains how this measured trust, as a part and parcel of a patient's experience of traditional medical practice, will support her (i.e. the patient's) acceptance of new technologies (in this case, prenatal screening for Down's syndrome) even though the patient may have difficulties understanding the results of these technologies:

> If I go to a doctor and the doctor says well your gall bladder should be operated … it is very rare that I go for a second opinion … I believe the first one. He's the doctor and he told me to go for an operation so I go. So all of this [screening] is the same thing … that we have the trust. I don't know what the level of information [is] and how well people really do understand. For instance, the ideas … the philosophy behind these different screenings [which are] totally different …
>
> (F 5)

When pregnant bodies undergo these invasive tests, this austere self-disciplining of reproductive asceticism can be viewed and experienced as necessary for the overall, external regulation of 'fit' populations in consumer culture. In this regime, the female body emerges as a reproductive resource; a gendered but nevertheless corporeal asset through which she can assume an indelible moral identity (Tietjens Meyers 2001). Drawing on Bourdieu's work, Shilling (1993: 127) notes that in contemporary society, the body has become a more comprehensive form of physical capital, a possessor of power, status and distinctive symbolic forms – integral to the accumulation of resources.

Specifically, in reproductive genetics, female bodies can be seen as becoming women's physical capital. In turn, women's physical capital becomes inexorably linked with their genetic capital, as they encounter the regulatory practices of reproductive genetics.

The production of physical capital and in turn, genetic capital, refers to the development of bodies in ways that are recognised as possessing value in a variety of social fields. Translated to the institution of reproduction, women's embodied participation is converted into caring and consuming for her foetus. Pregnant bodies become communicative bodies (Frank 1991) deeply engaged in the process of producing life.

From this viewpoint, genetic capital is culturally constructed as being 'in bodies' and providing genetic status. Here, I want to take a slight diversion in my discussion of genetic capital.

Donna Haraway (1991: 39–40) uses the term 'genetic capital' when she describes the work of two sociobiologists who work within the 'population genetic' developments of their field. For them, the past sets the rules for the potential future and biology creates a system favouring traditional gender roles in stable systems analysed with functionalist concepts. Their version of genetic capital appears alongside natural selection. Genetic capital is all about appropriating the labour processes needed for survival as opposed to extinction. Genetic capital is all about people involved in human evolutionary practices.

The sociobiological version of genetic capital is very different from the sociological one. Viewing genetic capital as biologically rooted, as sociobiologists do, is too limiting. We need to see how one's biogenetic endowment and the cultural representation of genes 'in bodies' are linked in a definition of genetic capital. Genetic capital should be perceived as that physical capital which functions to establish one's genetic identity and status in society.

That the sociobiologists' ideas on genetic capital are linked with their ideas on developments in technology is an interesting point noted by Haraway. Sociobiologists contend that how our human bodies transmit genes has not kept pace with cultural transmission of technology (Haraway 1991: 35). In other words, sociobiologists believe that our system of production including developments in technology has transcended us (Haraway 1991: 35). They uphold the idea that man's [sic] biogenetic endowment (i.e. genes) is obsolete and has not adapted to accelerating technological processes. Thus, they argue that biology has been transformed from a science of sexual organisms to a science reproducing genetic assemblages (Haraway 1991: 45). Their viewpoint emphasises how biologenetic endowment can be challenged by technological progress as well as scientific knowledge of and manipulation of genes.

In this context of accelerating technological processes, Marilyn Strathern (1999: 209) adds to this discussion of genetic capital. She looks at how procreative bodies, focusing on the 'generative moment' are adapting to new technological processes in assisted reproduction, specifically through the NRTs. For her, human biogenetic endowment is being privileged, as NRTs are attending to biological connections and I would add, altering in some cases the transmission of one's genetic capital. Strathern contends that technologies of procreation enable one dimension of kinship to succeed – the extent to which people have always 'worked' to make kinship. While technology generally can have its enablements and empowerments for better or ill, she asserts that the key feature about technology is that it works, whether it is successful or not.

While prenatal technologies may not alter biological connections and the transmission of genetic capital, as the NRTs do, these technologies alter the embodied relationships between pregnant bodies and their foetuses. The embodied connection that pregnant women have with these technologies may shape what they think and feel not only about these particular invasive technologies but also the foetal bodies within their own bodies. While prenatal technologies do not always work, as we have seen with regards maternal serum screening, they are successful in transforming the deeply embodied experience of pregnancy into a series of technological events, which endorse good genetic capital.

In the feminised regimen of reproductive genetics, women enact a morality of the body which upholds the external population's standard – the desire for conventional (i.e. non-diseased, genetically normal) offspring and the need for citizens who are fit to be born. That women's physical capital becomes inexorably linked with their genetic capital exposes the triumph of reproductive genetics, controlling women's involvement in the genetics discourse. Genetic capital is profoundly embedded in women's experience of pregnancy. Genetic capital can be ranked with a variety of unhelpful consequences, as we shall see in the following chapter. As pregnant bodies are inscribed by a powerful discourse on genetics in and through the regulatory practices of reproductive genetics, genetic capital not only determines whether or not a particular woman's body should be viewed as a reproductive resource but also is central to constructing a genetic moral order. Thus, when experts speak of 'affected offspring' or 'risk' and utilise technologies to rid wombs of 'non-viable foetuses', they are actively supporting the population's desire and society's supposed need for fit bodies as well as firmly establishing the link between physical and genetic capital.

Here, I want to show briefly how a sample of expert research findings surrounding the proliferation, use and evaluation of prenatal
technologies, especially prenatal screening, tend to be shaped by population, financial, technical and service delivery concerns, rather than
embodied ones. These sorts of concerns suggest that pregnant women
are merely disembodied means to an end (i.e. the production of fit
bodies). For example, physicians argue that prenatal technologies must
maximise the net benefit to society (i.e. abortion of defective foetuses
should be balanced by an acceptable false positive rate) (Beazoglou *et
al.*1998); the cost per case of detection of a disability should be less
than the cost per case prevented (Cunningham and Tompkinson 1999);
pregnant women should prefer a procedure related miscarriage than
birth of a Down's syndrome child (Kuppermman *et al.* 2000); prenatal
screening services can be provided in a 'one stop' multidisciplinary clinic
(Spencer *et al.* 2000); prenatal screening should be available to all women
(Spencer 1999a) and invasive procedures' (i.e. CVS and amniocentesis)
complication rates do not equal patients' (i.e. pregnant women's)
perception of risk of Down's syndrome (O'Connell *et al.* 2000).

Interestingly enough, when experts reveal some of the negative
consequences of prenatal technologies and focus their attention on
problem areas, they shape their concerns primarily around procedural
related concerns. For example, maternal serum screening is a procedure that is not universally acceptable but prenatal genetic testing
could be (Verlinsky *et al.* 1997). Anxiety related to a screen positive
result probably causes decreased participation in the next pregnancy;
the false positive rate must be reduced (Rausch *et al.* 2000). There
exists confusion about the reliability of risk estimates produced in
Down's syndrome screening programmes and many centres are unsure
about what methods of calculating risk and what population parame-
ters should be used; they must produce more reliable calculations
(Spencer 1999b).

Regardless of whether or not physicians are aware of the negative
consequences of these procedures and shape their research and other
interests through disembodied concerns, they employ prenatal technologies in order to purge pregnant bodies of abnormal foetuses,
reinforcing the link between physical and genetic capital.

Reflecting critically on the proliferation of these prenatal technologies, one scholar, Ruth Hubbard (1985), has suggested that it is not so
different to abort a foetus who is a girl in China and a Down's
Syndrome one in the USA: both decisions are deeply embedded in
social pressures. In a related context, the same author (Hubbard 1986)
refers to these pressurised decisions as 'obscene choices', confronting

pregnant women. Here, I would add that these decisions become not only obscene but also embodied choices for pregnant women. One clinical geneticist contextualises the pressures and the lack of freedom pregnant women and their partners have:

> If the society is discriminating against sick people and handicapped people then it reflects on the individual, personal and familial level too – that people who otherwise would and should be free to make their choices about their children. For example, if they want to have a handicapped child, they do not feel free to make their choice if the general attitude in society is that we are trying to get rid of the handicapped people and they are just useless people and that it is better not to have handicapped people.
>
> (F 4)

Side by side society's standard for reproducing able-bodied citizens exists a powerful representation of the female body as a vehicle for women to achieve motherhood or a reproductive life style. An extreme version of reproductive asceticism not only upholds the body as machine view but also represents the female body as a commodity and 'children' themselves as consumer objects – subject to quality control.

As we have seen from the above discussions, the body as constructed by reproductive genetics is made to play down the importance of human agency, particularly, embodied female agency. Already within the institution of reproduction, pregnant bodies have been inscribed 'as passive' through a variety of biomedical practices: her uterus is active, but its activity is not a subject's activity, it is impersonal and mechanical (Sbisa 1996: 368). Specifically, the embodied experience of reproductive genetics brings with it the imposition of a fragmented morality of the body with regards female moral agency. While pregnant bodies have been inscribed as passive, a major element in the discourse of reproductive genetics is that good performing pregnant bodies should enlist in active pursuits concerning knowledge and expressions of their genetic capital. If not, these bodies are seen as failures in the genetics moral order. Performing the pregnant body correctly in reproductive genetics is not only about self-discipline but also about incitements to a precise moral posture.

5 Gendered bodies, the discourse of shame and 'disablism'

> One level [of awareness] concerns the psyche: how a woman perceives her body – which is hard to separate from who controls it – which, in turn, helps to determine how the controls can be altered.
>
> Ellen Frankfort, *Vaginal Politics*, p. xxviii

Introduction

As in the previous chapter, I continue in this chapter to develop a feminist, embodied approach to reproductive genetics. In my explorations of the complexities of reproductive genetics and my desire to establish gender and the body as key theoretical sites, I want to discuss some of the social context and value issues in reproductive genetics framed by the concepts of the body, especially 'the good body'. My main contention is that women bear the burden of reproduction per se as well as the production of valued, healthy, non-disabled offspring.

Landsman (1998) suggests we live in the 'age of perfect babies' and as a result, mothers of disabled children are seen as producers of 'defective merchandise'; their pregnant bodies, as embodiments of motherhood, have failed to follow the culturally appropriate trajectory. To avoid negative judgements about the 'products' of female embodiment, pregnant women must engage with prenatal technologies in order to ensure that they have the correct genetic capital and 'good enough' reproductive bodies. Here, I would contend that pregnant women's sense of embodiment mediates their experience of participation in regimes of prenatal technology, similar to other technologically screened women who encounter other forms of medical scrutiny (Howson 1998; Bush 2000).

The attention on the workings of the female body in reproduction; the 'good' female body as a valued 'foetal environment' and the links between the pregnant body, disability and reproductive genetics are implicit issues which will be dealt with in this chapter. Moira Gatens (1992) has argued that we must attempt to 'write' the repressed side of the body and to disarrange the discourses in which dualisms operate. Thus, in this chapter, I will attempt to challenge the 'conspiracy of normality' (Stronach and Allan 1999), embedded in the disability discourse which pregnant bodies must confront. The separation of disability from normality needs to be contested, especially with regards the use of prenatal technologies.

As Bailey (1996) contends, prenatal testing has institutionalised the fear of impairment and increases the values attached to normal, non-disabled children. While the non-disabled body is often contrasted with the disabled body, pregnant women become acutely aware of this contrast as their own bodies and foetuses are shaped by their encounters with technology. Whether or not a pregnant woman chooses a selective abortion to rid her body of a potentially disabled child, this, as we have seen, is an embodied choice: her choices are constructed within the context of constraining discourses, shaping her pregnant body. Furthermore, any pregnant woman's embodied assessment of whether her foetus's (the potential child's) life is worth living can only be based on what her life and body means to her (Morris 1991: 69). While most, if not all, medical professionals working in the area speak about maximising choice (Lilford and Thornton 1996), the medical functional approach to disability tends to emphasise quality assurance (Priestley 1995) and good results.

Besides the discourse about disability and disabled bodies, there is another discourse into which pregnant women are further captured – the discourse on shame. That the experience of pregnancy *vis-à-vis* reproductive genetics has the potential to cause divisions between women on the basis of their genetic capital and reproductive potential is disheartening. While reproductive genetics focuses attention on attempts to contain if not eliminate bodily diversity and to standardise an ideal body, genetic notions place disabled bodies out of desired contexts – in work, living independently, in the public gaze, and so on.

In this chapter, I will look firstly at how the discourse of reproductive genetics makes divisions amongst pregnant women who emerge as either 'good reproducing bodies' or 'bad reproducing bodies'. Secondly, I want to illustrate the social effects and limitations of reproductive genetics in relation to disability as a cultural representation of impaired bodies. The final discussion in this chapter looks at how society attempts to place disabled bodies in social spaces marked by isolation, separation and exclusion.

The pregnant body and the discourse on shame

As a powerful way of injecting biology and hierarchy into social relation-ships, reproductive genetics constructs the idea that genetic capital, pedigree (i.e. pure breeding) and ultimately, social fitness can be ranked. But, of course, this ranking is carried out in already unequal social contexts in which gender and disability are devalued. Caught between bodies ranked according to their pure breeding potential and those ranked according to discursive systems of inequality (i.e. class, race, gender, etc.), pregnant bodies can be disciplined further by a discourse on shame. Here, shame is about an awareness of some serious flaw or mistake in oneself: one is guilty because one has committed a wrongdoing.

Sandra Lee Bartky (1990: 84) has argued that certain patterns of feelings are gender related and that the peculiar dialectic of shame and pride in women's embodiment is consequent upon a narcissistic assumption of the body as spectacle. Thus, with regards this notion of the body as spectacle, shame always needs 'the Other' – an audience (Bartky 1990: 85). While Bartky contends that women are more prone to shame than men, shame is a profound mode of disclosure both of the self and one's situation. For pregnant women with 'bad' genetic capital, expressions of their shame are stylised revelations of 'not good enough' bodies and 'poor' procreative capabilities within the institu-tion of reproduction. Their shame is an embodied failure to reproduce successfully. This 'pregnant' shame, similar to shame generally, is a pervasive affective attunement to the social environment (Bartky 1990: 85) which values 'good' and 'normal', that is non-disabled bodies over disabled bodies.

There has been work done on the gendering of shame as an emotion and how women have been made to judge and evaluate their bodies according to the ideal image and standards of the female body as 'flaw-less' or 'perfect'. Many women are involved in bodily regimes of intense self-monitoring and discipline (Urla and Swedlund 1995) in order to fit into this feminine image and uphold these heterosexist standards. Perfect bodies or beautiful bodies are given more physical space in our social worlds (Lorber and Yancey Martin 1997). What we do know is that some women have rejected the media or Hollywood images of the perfect female body (Morris 1991). This image of the perfect female body has been rejected by some men and women because they believe that outward appearances are not the only way of judging the value of a woman. Their message is: 'It's what's inside that matters'.

In a somewhat invisible way, reproductive genetics has succeeded in elaborating this message, but re-working it as 'It's what's inside *bodies* that matters'. This is because through reproductive genetics, experts are saying 'what's inside bodies' not only matters but also what's inside bodies (i.e. genes, foetuses and wombs) can be valued and ranked in far reaching ways. In a very real sense, pure breeding, good lines in one's pedigree and genetic harmony are privileged in the formation of genetic capital. In this system of ranking, pregnant women are divided into two opposite types of bodies or foetal containers on the basis of how well they reproduce.

In this context, Thalia Dragonas (2001) carried out an illuminating study of pregnant Greek women with 'risky' prenatal screening results and invitations to undergo amniocentesis. She found that these pregnant women had a sense of responsibility and accountability for a potential 'faulty embryo' and hence, they experienced shame and guilt. She argues that shame was readily available to pregnant women because pregnancy concerns them in a more visceral, direct fashion than it does men. Furthermore, society, by rewarding motherhood, allocates to pregnant women all profits and losses in producing and bringing up children (Dragonas 2001: 142). She contends that the sort of shame that these women experience is about the self – about the fact that the self as mother has a supreme transgenerational mission, a major duty that has to be fulfilled to secure family lineage. When 'faulty foetuses' are found, women see themselves as failures. This sort of self-debasement places the vulnerable pregnant woman in a 'low position', while the strategy of shame emerges from 'the phantasy of a menaced foetus' (Dragonas 2001: 143).

Thus, the view emerges that on the one hand, there are those women who are good reproducers with sound wombs capable of breeding well and with healthy genes. They have good genetic capital. They should reproduce children and make babies. On the other hand, there is a group of women who are bad reproducers with problematic wombs not capable of breeding well and with faulty genes (Finger 1990). These women have the potential to reproduce genetic mistakes or bring into the world those viewed as not fit to be born. Their bodies are flawed and they should not reproduce or make babies.

Having social worth in reproduction, good genes or valuable genetic capital dictate the boundaries of the genetic moral order. Within this moral order, genetic harmony, unity and compatibility between a foetus and a pregnant woman become important in so far as these notions imply fitness, good genes, viability and health. Pregnant women, experiencing the opposite, confront a state of disharmony and potential ill health. This is an unintended social consequence of reproductive genetics, creating divisions between good reproducing bodies and bad reproducing bodies. Simply, women are separated into good reproducers and bad reproducers. Thus, the idea that first-rate and second-rate female reproducers exist is propagated and emphasised.

As the technologies of reproductive genetics invade pregnant women's wombs, some if not most, regardless of their state of genetic harmony, endure their pregnancies as times of anxiety and stress (Katz Rothman 1994). Furthermore, women who are bad reproducers inevitably experience psychic distress (Farrant 1985). One leading expert in the field discussed this type of anxiety by referring to a recent headline:

> *Screening test designed to allay anxiety has reverse effect* … What does that headline tell us? Is the screening test designed to allay anxiety? That journalist didn't stop to think. Of course, he [sic] didn't. A screening test creates anxiety in a sense, it's deliberately doing that.
>
> (Epidemiologist E 6)

For women with 'bad news' and reproducing 'genetically bad' children, shame becomes linked up with the production of individuals not fit to be born. In this context, it is interesting to see how experts use value-laden language when they speak about the use of prenatal technologies. They demonstrate that in the social relations of reproductive genetics, moral judgements, endorsing the 'good' not 'bad' foetuses and implicitly, good and bad reproducing bodies are made. For example, experts' use of the terms, 'bad news' or 'bad results' was a refrain often made when they were making claims about the significance of prenatal technologies. Here, the epidemiologist, who spoke above, talks now about the 'bad news' one receives when screening is positive. He reveals some of the subtleties of prenatal screening. Below, he is referring to the triple test:

> So I am going to tap you on the shoulder and give you bad news. You are going to be extremely distressed by that bad news. But there is an important consequence in the news that I gave you and you have a number of options that you can take – some of which … you may wish to take recognising that you are going to be quite upset. What's important is that people have a choice as to whether [or not] to get into that situation in the first place. But, once they are in that situation, they will either be more or less where they were before, that's of having no news … Screening negative does

not mean you won't have bad news. That's another question. It means that you have got a lower risk of Down's.

(E 6)

Another expert, a clinical geneticist, views receiving a 'bad result' as health threatening and he compares receiving a 'bad result' with hearing that 'you are going to get cancer':

> [Pregnant women] take the test ... they get the good result ... they can forget it ... The few who get the bad result will get ... information ... If I go to a dentist and he says there is a cavity – I am not badly disappointed. Just like you hear you're going to get cancer and you say: 'Well it is just good to know'.
>
> (F 5)

Defending the use of amniocentesis, one expert acknowledged that there can be 'bad results' and this was, in his view, a minor occurrence. He supports amniocentesis and mentions those people 'with strange stories in their families' who were unable to have the possibility of prenatal diagnosis. Interestingly enough, he denied medicalisation and the 'tentative pregnancy' according to the views of Barbara Katz Rothman (1994) whose work he was aware of:

> I don't support very [much] ... the ideas of Barbara Katz Rothman who says that you medicalise. You make it difficult. You have this so called tentative pregnancy ... It is of course that just being happy is being postponed perhaps. The other side of the picture is always that, I am old enough to know. Still, [there are] people who decided not to have offspring because there were strange stories in their families ... [They] now tell me that, if in their time, prenatal diagnosis were possible, they would have had children. People get a bad result from amniocentesis, but 99 per cent get a good result and are happy. It's like being worried that they are 38 or ... that they might get a Down's syndrome child.
>
> (Gynaecologist NL 2)

In this context, I would contend that a new type of social wrong-doing or transgression becomes visible for pregnant women: the bad reproducer. If this bad reproducer decides to continue with a pregnancy with a 'bad result', she may experience further shame as one who decides to reproduce badly. Given that there is no perspective which recognises the capacity of the individual woman to negotiate their own lives within the constraints of genetic diseases (Parsons and Atkinson 1993), a bad reproducer appears as one not only having a bad body, bad genes and worthless genetic capital, but also daring to allow 'bad genes' or 'genetic mistakes' come into this world. She appears as exerting no restraint, no discipline on her capacity for impure breeding. But, in a real sense, by deciding to bring a disabled body into the world, a pregnant woman may be challenging the notions of bad and good reproducers and the prominence of non-disabled bodies over disabled ones.

In the context of validating disabled people's lived experiences, James Overboe (1999) discusses Gilles Deleuze's call for 'a diagnoal approach' that recognises difference without negation. For Overboe, a diagonal approach allows disabled embodiment and sensibility to be perceived as a valid way of life without being negated. This diagonal approach does not negate disabled bodies rather disabled bodies are recognised as different from other bodies. He contends that in this way, disabled embodiment is perceived as one way of being in the world. Thus, if pregnant women choose to have disabled babies, they can be seen to be upholding this diagonal approach.

As we can see from the above discussion, the concepts and ideas utilised in reproductive genetics appear to exert more restraint and impose more limitations on women's than men's bodies. Here, the construction of the body has been the effect of endless circulation of power and knowledge (Bordo 1993a: 21). This body, as gendered, raced, aged and marked by anatomical, cerebral or physiological difference or damage, provides the focus for regulatory techniques which are practised on the individual body, particularly a pregnant body, as a living body. These regulatory techniques are continually being played out through reproductive genetics and reproductive health care within the institution of reproduction. High-technology medicine (Williams 1998) can be seen to affect the mindful, emotionally expressive body – the pregnant woman. At this point, the external problem of regulating a consuming body and the internal problem of exerting self-restraint on a reproducing body merge with the result that shame, as an experience bound by gender, performs a regulatory function. The morality of the body during pregnancy demands order or genetic harmony. For pregnant bodies, 'fit or viable foetuses' means they have fit bodies and they can share in this harmony. While unfit foetuses may imply states of disharmony and potential ill health, the self-restraint practised through their reproductive asceticism comes to be seen as a failure in the eyes of these bad reproducing bodies. Similar to women who have difficulty in having children (Exley and Letherby

2001), pregnant women whose foetuses are 'not viable' are likely to experience disruptions in their lifecourse expectations (i.e. to be a mother). Thus, these disruptions may compel them, as bad reproducers, to do 'emotion work' to reaffirm their identity as women.

Reproductive genetics and disability

While the public may have a favourable attitude towards the general availability of prenatal technologies in reproductive genetics, a frequently articulated response among some groups of disabled people is that test results might lead to some form of social discrimination (Shakespeare 1995; Gillam 1999; Bricher 1999). Chandler and Smith (1998) argue that not only do these techniques have discriminatory effects on disabled people but society risks promoting 'a culture of perfectionism' in trying to eradicate disability through these type of technologies.

In this context, there is a growing interest in disability studies and some excellent and thorough going accounts concerning the social model of disability (Barnes and Mercer 1996; Barton 1996; Hales 1996; Oliver 1996; Shakespeare 1995; Barnes 1992; Wendell 1996; Corker and French 1999). Upholding this social model, Kelly and Field (1996) contend that medical sociology has understated the central facts of bodily difficulties, entailed in illness and disability. Within the social model of disability, two notions, disability and disablism, have been defined by Carol Thomas (1997: 623) and are illuminating in this context:

> *Disability* is not the condition or functional consequence of being physically or mentally impaired. Rather, dis-ability refers to the disadvantaging affects – referred to by many – as the 'social barriers' – faced by people with impairments flowing from *disablism*: the ideological antipathy to what is considered to be undesirable physical, sensory or mentally related difference or 'abnormality' in western culture (my emphasis).

Inherent in the concept of disability and the process of disablism is the idea that a pervasive medically describable paradigm of human, physical ability or mastery of the body is possible (Wendell 1992). As Tom Shakespeare (1995: 24) aptly states: 'People are disabled not by their bodies but by society'. From a similar viewpoint, Wendell (1992: 70) notes that 'the oppression of disabled people is the oppression of everyone's real body', meaning that regardless of the fact that bodies are tremendously 'diverse in size, shape, colour, texture, structure, function, range and habits of movement, and development', this is not reflected in our culture. Wendell contends that even though physical ideals change over time, we romanticise the human body. Cultural ideals are not just about appearance, images or looks, but also these ideals are about human comportment. Cultural ideals on the body are concerned with how well bodies confront and conform to delineated social spaces.

Disabled bodies, 'disablism' and reproductive genetics

With these cultural ideals, we make judgements about proper and improper bearings, posture, carriage and mien of bodies. Bodies, inscribed by disability, displace the cultural ideal of the positive, coping, well-situated, fine-carriaged, erect and proper social body. As a morality of the body, disablism aims at disciplining bodies in the social arena. It emerges as a fundamental piece of the aestheticisation of life (Featherstone 1991: 67); disabled bodies are viewed as disrupting the aesthetics and artistic sense of the ideal body in our consumer culture. While there may be disability inflected modes of embodiment (i.e. different types of disabilities and ways of controlling disability) (Hughes and Witz 1997), disabled people are viewed as those who disturb conceptions of normal embodiment. The biomedical control and indeed elimination of potential disabled bodies by prenatal technologies is all about pregnant women being forced into confronting a disability inflected mode of embodiment. DNA-based prenatal technologies are used 'on bodies' to detect foetal abnormalities with the effect that the reproductive process is medicalised further (Birke *et al.* 1992: 154).

As genes and bodies are being culturally shaped, pregnant and disabled bodies are meant to fall outside of the paradigm of health or well being as well as the universal, medically describable paradigm of human physical ability. The cultural prerogatives (i.e. to have perfect children) directed towards women during pregnancy are employed strategically by the medical profession to engage them in the disablism discourse. Most, if not all, pregnant women are meant to have an 'antipathy to what is considered to be undesirable physical, sensory or mentally – related difference or 'abnormality' in 'their bodies' in Western culture. Two experts mirrored this view:

> [When] I think always [of] pregnant women, there [is] no pregnant woman who wants to have a handicapped (sic) child. They all want to have a healthy child …
>
> (Gynaecologist NL 6)

and

> If you say to a woman that your baby is a Down's, what do you think? Do you think that she is going to terminate the pregnancy or not? In Greece, she is going to terminate the pregnancy.
>
> (Obstetrician G 3)

Another expert, a gynaecologist, exposed his own discomfort for disabled bodies when he said bluntly:

> Setting out to have an accident is different from preventing it ...
>
> (NL 2)

In a related context, one medical geneticist's wants, reflected what appeared to be the wants of pregnant women (i.e. to have a healthy child):

> Personally, of course, I am against [interrupting] a pregnancy except if the child is going to be a handicapped child.
>
> (G 5)

Another expert (a lawyer) extended these wants to the wider society:

> For the general public ... they want healthy babies and [want to] get rid of [the] disabled.
>
> (G 9)

What should be flagged up here is how in the engagement between reproductive genetics and the bodies of pregnant women, the genes, appearing as most significant, are those linked with what is perceived as 'disability' and, in some countries, sex (Wertz and Fletcher 1993). This engagement is linked to the wider cultural debates and regulatory practices on the meaning of bodily differences. Nelkin and Lindee (1995b: 399) argue that 'genes appear to cause bodily differences that matter most in these cultural debates'. They contend further that 'bodily difference is historically specific, written not in the body, but in the culture that defines what aspects of the body are most important when one begins to sort people into groups' (Nelkin and Lindee 1995b: 399–400).

While reproductive genetics focuses attention on attempts to contain if not eliminate bodily diversity and to standardise an ideal body, genetic notions place disabled bodies in desired contexts – out of work (Barnes 1992; Pinder 1995; Oliver and Barnes 1998; Abberley

1996), in care (Chappell 1994; Drake 1996), dependent on others (Morris 1991; Keith and Morris 1995), not reproductive (Finger 1990) and so on, and exclude them from other contexts – in gainful employment, living autonomously, in the community, etc. One expert contextualised the social exclusion of disabled people:

> You don't see special spaces for wheel chairs especially in the [countryside] villages. Disabled people used to be secret – something that should be hidden from society – something that people were ashamed of. Of course, these attitudes are changing now but still I think we are quite behind in this area.
>
> (Lawyer G 7)

Disablism: from state benefits to benevolent technologies

In a real sense, society attempts to place disabled bodies in social spaces marked by isolation, separation and exclusion. Regardless of the fact that this sort of marking may be gradually changing, emergent expert discourses come to own the problem of disability through the creation of powerful discourses about that 'problem' (Hughes 1998: 75). Nevertheless, this process of excluding disabled bodies from wider society tends to be legitimated by the fact that disabled people, as a group of 'non-aesthetic bodies' and 'unattractive citizens', are seen to benefit from the benevolence of the welfare state. Implicit here is a type of 'state sponsored model of disability', promoting individual failing above any attention to environmental or social factors (Shildrick 1997: 57). In reproductive genetics, the existing logic is that the genetic moral order (i.e. improving genetic capital and, ultimately, bodies) and its attendant technologies are able to proliferate in society as long as the state provides a certain 'high' standard of disability benefits (i.e. taking good care of disabled people). The following two experts reflect this type of logic:

> What is good in the whole of Western Europe is that at the same time we are developing the testing and the screening for genetic diseases, we are providing genetic services. We are improving the race ... and in a way, we are also taking quite good care of handicapped people ... this shows that our society has a moral[ity] which is still good ... As long as we are taking good care of those with diseases and [the] handicapped, then we can trust that it [technology] is not misused. But ... I am afraid that it is possible

to discriminate [against] more and more sick people and handicapped people. It is quite obvious that it is happening already.

(Clinical geneticist F 4)

and

I think, that if we are going to have prenatal or genetic screening
generally, it is important at the same time to have a commitment
to continue services for people who are disabled, because otherwise ... they will have less rights.

(Policy maker E 8)

Nevertheless, as disabled bodies become progressively de-valued
and disablism shapes the dominant discourse on the body in reproductive genetics, benevolence in the form of benefits from the State will be
transformed to benefits received from the technologies themselves.

Simply, technologies used to abort, for example, Down's foetuses
will be increasingly viewed as helping 'potential' disabled people
and/or their 'relatives'. Here, it is clear that there is little, if any need
for experts to hear the voice for disabled people in the diffusion of
prenatal technologies (Bricher 1999). One expert expressed how the
idea that the technologies themselves are helping relatives to be (i.e.
future parents) and that this 'benevolent' process is not new:

I do not know much about the opinions within the group of
handicapped people, but I know there is some discussion about
prenatal screening because it would mean, we do not except any
longer handicapped people. I don't think that's the case, because
it's not a denial of handicapped people. The only thing you want
to reach with prenatal screening is to in fact ... to help parents to
be of handicapped people ... Some of them don't want to have a
handicapped child. I don't think that's much different now [than]
from the techniques we had 30 years ago. Then ... we had some
techniques to look at the unborn child and now ... the techniques
changed, but not ... what we think about and what people think
about.

(Gynaecologist NL 6)

Another gynaecologist expressed a similar view, but suggested how
this type of medical benevolence was welcomed by the disabled people
themselves:

I always find it ... [with] people with neural tube defects ... they come themselves to [me to] have amnionic fluid tested for alpha beta protein because they want to know if their baby will have a neural tube defect or not. So these people might be the strongest advocates against prenatal diagnosis but, no, they are often the strongest supporters ... We ... have our chairman of the [Local Disabled Society] ... she has a child herself and she is our strongest supporter for prenatal testing.

(G 2)

In a related context, early research (Breslau 1987) confirms that mothers of disabled children are in favour of prenatal testing. Indeed, on the one hand, disabled women or mothers of disabled children often share in the wider social and medical discourses on what constitutes reproductive risks. On the other hand, considerations of risk of 'foetal' disability becomes an important feature for a pregnant woman's, whether able bodied or disabled, reproductive journey (Thomas 1997).

Uncomfortable ethics and disabilism as the dominant morality of the body

Regardless of the fact that prenatal technologies are also seen as benevolent and improving the condition of disabled people, reduction of birth incidence of genetically flawed individuals may place society on an uncomfortable ethical landscape (Davison *et al.* 1994). In this context, one expert believed that he had already been placed upon an uncomfortable ethical landscape when he said:

You can't ... be held to ransom in a way by the people who have ... disabilities ... because you do this [in this instance prenatal genetic testing], [they] ... don't like us.

(Medical geneticist E 3)

One researcher (Ellison 1990) asks: 'Will society ever change so far as to question the birth of children with severe inherited disease and raise doubts about the responsibility of parents who fail to avoid such a situation?' Related to this question, the following expert, an ethicist, reflected on how the views towards disabled people may be changing in society as a result of the use of prenatal technologies:

It is a worry that a lot of people have ... it might come ... through [in] attitudes towards women who have chosen not to terminate when they have had prenatal screening. And that might eventually feed through into a view that they made a wrong decision. And that this disability was avoidable, and so therefore society shouldn't have to pay for them. That's a big fear. I don't think that it has happened yet but I think it could do … But I think it will be the case that people will become less tolerant.

(E 9)

Another expert reflects on the psychological implications of these technologies on society and how the expansion of these technologies relate to views on disabled people:

We've heard a lot of people saying: well it's just about eliminating illnesses and suffering. It is. But, what psychologically happens in society if that is the way we are doing it? And then last but not least … [Is this] … where the energy needs to go? How does expansion of screening relate to the way society views … disabled people? And that to me is where we need to put our energy. So I think if we look at the disability implications with that, that will be a useful way to address some of these problems.

(Researcher E 1)

While giving birth to a disabled child can be seen as irresponsible, pregnant women's embodied choices need to be given moral significance and they need to be treated with respect regardless of the reproductive options that they choose (Purdy 1999: 128). One expert, a gynaecologist, was keenly aware of the ethical dilemmas involved in the use of prenatal technologies for eliminating potentially disabled people and he felt that it was unethical to consult disabled people:

It is rather unethical to ask someone who is disabled what he [sic] thinks about prenatal diagnosis.

(Gynaecologist NL 2)

Regardless of the above, the state will more than likely continue to uphold disablism as its dominant morality of the body in order to serve the interests of scientists with their emergent prenatal technologies, the medical profession and the general public. While the State may need these self same genetics experts (i.e. reporting here) to guide government and society in decision making about the new genetics (Kerr *et al.* 1997), the State also needs 'able bodied citizens' as mothers, workers and warriors. Here, an inclusive debate is needed which includes professionals, government representatives and lay people, including representatives from disabled people's groups (Kerr *et al.* 1998). The State also has clear economic concerns, as expressed by the views of the following experts:

> This is very ... expensive to raise all these disabilities [sic] disabled children.
>
> (Psychologist NL 7)

and

> If you have a disabled child, there are practical issues concerned with his [sic] education, his [sic] position in the family, the effect of such a birth on other members of the family and the other children, the financial cost of raising such a child especially in [country] where such institutions are not very many.
>
> (Lawyer G7)

and

> The child may need an extra care. Down's children often need extra [care] and they need lots of day care systems and things, which cost money. And we have the recession ... and we have this less and less tolerant climate ... I mean, it has happened already in a number of countries at least it is happening in [European country] and I think it probably has in the [North American country], [It's] about people actually ... pressurised not to give birth to children with disability because it will be costly. And pressurised in a way that they then have to pay the cost themselves, if they want to give birth to this child although they know it's going to be handicapped.
>
> (Clinical geneticist F 17)

As we have seen from the above, powerful discourses and technologies are mobilised by experts on pregnant bodies. Regardless of any visible biomedical resistance to self-governing motherhood, the experience of reproductive genetics as mediated by experts has a profound effect upon pregnant bodies, women's reproduction and health care needs. Making babies or providing good stock can become for some, if not most, pregnant women a disembodied and disembodying process through which the foetal body becomes the centre of attention.

Both this and the previous chapter have generated an awareness of some of these unintended consequences of reproductive genetics by offering a feminist analysis of issues that become visible as 'the body' becomes a theoretical site in reproductive genetics. I have attempted to demonstrate that feminist embodied theories are needed for understanding the lived experiences of pregnant women within reproductive genetics. We have looked at the ways in which 'reproductive genetic concepts' are mobilised to mark 'harmony' or 'disharmony' in the progression of women's reproductive processes by privileging the mechanistic view of the body and placing reproductive limits on the female body through reproductive asceticism and a discourse on shame. The social effects and limitations of reproductive genetics in relation to disability and disablism were also discussed. The cultural prerogatives to have fit and non-diseased (i.e. genetically) children are directed towards women during pregnancy. These prerogatives along with attendant technologies are employed strategically by the medical profession to engage these women in disablism. Furthermore, as disablism shapes the dominant discourse on the body in reproductive genetics, benevolence in the form of state benefits are being transformed to benefits received from the technologies themselves.

Bodies, constructed by reproductive genetics, play down the importance of human agency. Bodies are ranked according to genetic capital – how well they conform to being and doing in a genetic moral order. Furthermore, society disciplines those 'potential' and 'real' bodies believed to 'contain' 'defective' genes more severely than those 'containing' 'normal' genes. This disciplinary process tends to offer a somewhat limited view of bodies' potential and shadows the fact that what may appear as 'defective genes' is in fact a result of the body's interaction not only with the environment but also gendered social practices. In this matrix, bodies appear to lack embodied agency, a quality that should be protected, valued and preserved in our contemporary society (Shilling 1998).

In future, more and more pregnant women, as female embodiments of the genetic moral order, may ask: 'Please doctor, may I have a normal baby?' If so, we need to provide an atmosphere in which an understanding of the social complexities and implications of asking that specific question becomes possible. In reproductive genetics, women's active agency and an awareness of bodies, inscribed by difference, need to replace the medical and social empowerment of prenatal technologies as strategic engagements in disembodying discourses.

6 Synchronising pregnant bodies and marking reproductive time

Comparing experts' claims in Greece, the Netherlands, England and Finland

> All the surrendering devotion our children have put into their private families these women have put into their country and race. All the loyalty and service men expect of wives, they gave, not singly to men, but collectively to one another ...
>
> Charlotte Perkins Gilman, *Herland*, p. 95

Introduction

In this chapter, I make cross-cultural comparisons of experts' claims from four European countries included in the experts' study: Greece, the Netherlands, England and Finland. Earlier work (Ettorre 2001b) on this study reviews cultural differences in terms of the countries' infra-structures for the development of prenatal screening; the implications of selective abortion; inconsistencies in service delivery and the issue of ethics. In this chapter, cross-cultural comparisons are made in order to illustrate key issues that become visible in a feminist and sociological, embodied analysis of the connections between reproductive genetics, gendered bodies, experts' knowledge claims, space and time.

Reproductive genetics relies on an individualistic, mechanistic view of a pregnant body not only marked by rigid definitions of gender and disability but also shaped by space and time. While the disciplinary practices of reproductive genetics are constructed by a variety of biomedical experts, these are produced in spaces, distinguished by over-arching cultural themes. These themes are played out in national spaces defined by racialised boundaries. In turn, these boundaries are marked out over time by nation states with geographical customs and ethnic traditions. These European spaces actively constitute visible, social and cultural practices, while at the same time, these practices shape and re-shape space. National spaces are made locally through social actions and power relations, making up the rules of particular societies and delineating state boundaries (McDowell 1999; Duster 1981).

Within these national, specific spaces, reproduction emerges as a social institution and matures into a regulatory system. As a corporeal style, pregnancy is framed by this social institution and combines with temporality and cultural spaces as a constituting and constituted context. Within the context of time-space analysis, Anthony Giddens (1981) defines institutions as structured social practices that have broad spatial and temporal extensions. In this sense, reproduction is a structured social practice that has a wide spatial and temporal expanse. Pregnant bodies are synchronised and harmonised by the technology of reproductive genetics within the institution of reproduction.

As we saw earlier in Chapter 3, in the movement from hospital to surveillance medicine, there is a shift in modern medical culture from looking at subjective signs and symptoms that sufferers, consumers, patients, pregnant women and others experience when they feel sick to performing objective tests in the community. With regards reproductive medicine, the consequences of this shift has meant that surveillance medicine's tactics (Armstrong 1995), pathologisation and vigilance, are played out on the bodies of pregnant women. These tactics accelerate the proliferation of genetic and other technologies into reproductive and foetal spaces. Here, there is nothing new in the medical profession's surveillance of these spaces (Oakley 1984; Newman 1996). What is relatively new are the medical construction of genetic risk factors for pregnant women; the technologies that allow detection of carriers and prenatal diagnosis of diseases and abnormalities, and pregnant women being made to experience reproductive time in a new way.

The over-medicalisation of pregnancy argument has been heard loudly and clearly in sociological circles for a number of years. Tentative pregnancies (Katz Rothman 1994) are temporal matters as well as temporally constituted. Physicians may ask questions about weeks of gestation and maternal age. However, for pregnant women, reproductive time takes on multiple dimensions beyond the nine-month timetable, when the consequences of their risks breach the boundaries between health and illness. Here, reproductive time can be seen to last a lifetime, if these women give birth to an abnormal baby.

Medical constructions of risk factors release reproductive spaces for consideration of future illness potential. Thus, the results of a nine-month pregnancy can be extended into another vista – one of illness potential. Cultural, reproductive and temporal spaces intertwine as co-ordinates in which risk factors are identified and choices are made.

In this surveillance process, pregnant women may focus on illness as continuous possibility, rather than on the immediacy and/or practicalities of conception. The future is now. For them, the impact of reproductive genetics has meant a transition from a naturally based manner of time to a socially and more specifically medically, based one. In this context, Margaret Stacey (1992: 2) says:

> Technological advances in reproductive medicine affect people's lives, endorse certain values, run into stereotypes and have consequences for the management of relations that may well extend beyond their immediate application ... To think of such technologies as having social dimensions provides a way of thinking about the multiple nature of their impact.

Here, thinking about the complex social dimensions of reproductive genetics allows us to consider the institution of reproduction both spatially and temporally. Reproductive genetics and the strategies of surveillance medicine are increasingly influencing reproduction. Reproduction embodies complex cultural, temporal and technological relationships and experts have the power to not only co-ordinate these relationships but also shape the level of involvement that pregnant women have with these complex relationships and processes. We may rightly ask how and what reproductive space and time are available to women and how are these culturally constituted. This chapter asks these important theoretical and empirical questions and flags up key issues for consideration in the four European countries mentioned above. In the first section, I will draw upon the claims of these experts. I will describe the specific cultural spaces in which they shape their claims about reproductive genetics and demonstrate how these claims mark reproductive time for pregnant women. In the second and last section, I will offer comparisons and conclusions.

Reproductive genetics within different cultures: space and time reckoning

A complex interplay between a series of biological, social and cultural processes and institutions such as medicine, science, gender and reproduction have standardised the way we currently reckon reproductive space and time. In looking at experts' claims about reproductive genetics, we are able to identify over-arching cultural themes country by country. These cultural themes include: thalassaemia in Greece, abortion in the Netherlands, ethics in England and care for pregnant women in Finland.

Greece's thalassaemia success story

Based on the Greek experts' claims, the thalassaemia success story emerged as the main cultural theme. The claim that biomedicine was successful in dealing with thalassaemia helped to shape the cultural space in which reproductive genetics was played out in Greece. It should be noted here that thalassaemia is a characteristic of the blood and carriers have an excess of red blood cells in comparison to non-carriers. Thalassaemia can cause severe anaemia because red blood cells cannot make enough haemoglobin. Treatment is periodic blood transfusions, but this causes high concentrations of iron in the carrier's organs. Dangerous concentrations of iron will be removed by Desferal infusions on an almost daily basis. For carriers, this treatment scenario can go on for the whole of their life span (up to 40–45 years of age).

In 1977, screening of thalassaemia began (Loukopoulos *et al.* 1990) because it was seen as a Greek disease, a public health problem. As a disease, it was viewed as living in the Greek population. Greeks perceived it as a stigma that marked carriers. For example:

> The person has in his [sic] blood this defect. He [sic] has a stigma ...
>
> (Lawyer G 4)

and

> The stigma shows that this disease runs in your family.
>
> (Policy maker G 1)

As a stigma, Greeks saw thalassaemia as having a high level of concentration in the corporal volume of its people. Thalassaemia became nationalised and enclosed culturally as Mediterranean anaemia. Here, one expert stresses this cultural boundedness of thalassaemia:

> We have ... taken certain measures that ... were positive in detecting Mediterranean anaemia because this is a problem in Greece.
>
> (Policy maker G 1)

While Greek experts claimed that the national screening programme for thalassaemia was a success story, success was possible only through a decrease in thalassaemia's wide, corporeal sweep across the population.

> We have had a great success ... because we used to have approximately more than 200 new cases per year and now [there] are less than 10 cases ... And that was really a big success ...
>
> (Medical geneticist G 5)

and

> We have a great problem with thalassaemia. There is an official laboratory ... for haemoglobin disorders ... because the incidence is high ... [Our programme] is 20 years old and the result ... is that we very rarely see now new cases – new children with thalassaemia.
>
> (Obstetrician G 3)

Experts allege that the incidence in the total population is high at 7 per cent to 7.5 per cent, and there are claims that there are special places that have higher incidences. For example, one expert says:

> There are special places that ... have higher incidence [higher than 7 per cent] and ... other places with lower incidence ... 7 per cent of the Greek population are carriers. That means that one in 14 couples are both carriers ...
>
> (Obstetrician G 3)

But, these exceptional places tend to be spatialised as rural areas:

> We have big pockets of thalassaemia especially in [the rural areas].
>
> (Policy maker G 1)

Given these big pockets, it is not only important to take positive measures but also to offer prenatal diagnosis and screening to couples before they marry. One medical geneticist claims that while Greece has approximately 100,000 live births per year, there are 6,000 pre-natal diagnoses. This meant that 6 per cent of the whole pregnant population are screened for chromosomal abnormalities. In this expert's eyes, this is a very positive effect of the national screening programme.

Experts claimed that nowadays physicians rarely see a patient with an affected baby and this is the result of everybody being part of the national screening programme or 'knowing before they get married':

> Now very rarely do you see a patient with affected baby who becomes sick ... All will have a prenatal diagnosis. Everybody is going for a test before having a baby and this is very important.
>
> (Obstetrician G 3)

and

> The best thing is to take measures ... and this can be done if [a couple] knows before they get married that they have both have this risk. This is being taken care of ... before marriage. We have a blood test and that's the way to do it because if you know what is coming to you, then your choices are yours ...
>
> (Policy maker G 1)

Another expert discusses the links between the national programme and prenatal diagnosis and how at risk couples are detected:

> [Thalassaemia] is a single gene disease ... In order to have an affected child, then two parents must have the trait ... The couple is at risk and each pregnancy carries the 25 per cent possibility of [having] an affected child. So ... we screen and detect couples at risk ... When we find a couple at risk, we try to offer the option of having the child only if they [sic] do antenatal diagnosis. [This is so that] if there is an affected child, they have to go for termination of pregnancy ...
>
> (Paediatrician G 6)

This expert also notes the care and time that is needed in this process. He claims that one is lucky to proceed towards a first trimester termination:

> We've managed to reduce considerably ... thalassaemia ... There [are] certain centres all over Greece that do screening for detection of couples at risk. Of course, we have had some [problems]. There were some false positives ... [We] screened and later [found] the [foetuses] were normal ... So you have to be very very careful ... [We

use] a blood test but you have to be experienced to do the diagnosis ... In these centres, the diagnosis is quite efficient. If we have
a couple at risk then there are organised units that we can refer
this couple to [for] prenatal diagnosis. And when you go to
prenatal diagnosis, we are dealing with haematological techniques.
The DNA technique is more accurate ... You can do it in the first
trimester of the gestation ... With the modern techniques, we are
capable [of doing] a diagnosis in 24 hours so you can know in
week 10–12 of gestation that you can have termination of the
pregnancy. So we proceed towards first trimester termination, we
are quite lucky.

(Paediatrician G 6)

As seen from the above, the thalassaemia success story shaped
experts claims about reproductive genetics. What is of interest is that
no expert mentioned pregnant women when speaking of these matters.
Pre-nuptial couples were their main concern and those with whom
experts would press for a reproductive timetable. Of course, timetables
prove to be important in clinical settings (Roth 1963).

The focus on couples was probably because both parents must
have the trait in order to have an affected child. On the other hand,
it is the bodies of pregnant women that are temporally and technologically co-ordinated specifically during weeks of gestation; not
those of their partners. Thus, it could be said that reproductive time
is employed to best advantage only when the space for gestation
(i.e. women's bodies and wombs) is over-looked. For Greek women,
this implies that while pregnant, their bodies become non-places
(McDowell 1999). Non-places are invisible spaces in which transac-
tions and interactions take place between already established customs
(i.e. male privilege) and institutions (i.e. medicine, gender, science and
reproduction). Thus, at any given moment in time, the perceived
wisdom about the spread of thalassaemia's corporeal volume in Greece
will determine the sorts of technologies that are practised on their
pregnant women.

The politics of abortion: control and liberation in the Netherlands

The main over-arching cultural theme from the experts' interviews in
the Netherlands was the politics of abortion. Experts claimed that
right wing Christians or the Christian Democratic Party (CDP)
featured centrally in political debates:

Policy makers ... are very much against large scale screening ... where the detection of an abnormality might lead to an abortion ... Because of these political conflicts that [the] government should support a screening programme [for] all women is a problem ... It ... is even a little bit of a problem to offer it to a specific age group...The reason for Parliament to pass this law against screening [is] ... that [screening] was seen as inadmissible ... because people (sic.) are already pregnant and [it is] inadmissible [as a] choice for abortion ... It's very interesting that the total number of the population that refuses abortion on any ground, is some 10–15 per cent ... Here, it was a political amplification effect of the right wing Christians, who could ... have their say on this subject. [They] were ... achieving influence [on] what was absolutely non-proportional to the number of the people in the population having the same views ... So it's very political ... Sometimes genetic abortions are ... taken as political ... and handled irrespective of the needs of the people involved.

(Clinical geneticist NL 4)

and

We have a CDP ... Up to three years ago, they were always in the government for 70–80 years ... They had to talk about [legal] abortion which was forbidden ... nobody wants to talk about it because you will get ... problems within your coalition in the government. Now that they have gone away from the government ... [the ruling party] are afraid that CDP will make a point of it opposition ... It's a very strange thing because ... we have a rather liberal abortion law, but if we talk about abortion because of genetic problems then there is a problem.

(Obstetrician NL 6)

Another expert said quite simply:

The option of pregnancy termination is not a justifiable reason for a population screening programme.

(Clinical geneticist NL 4)

In a similar context, another expert says:

[The government] doesn't like the abortion issue ... As long as we don't have any other treatment than termination of pregnancy, they don't like it as a political issue ... [Those] who have decided to legalise abortion to the 24th week actually don't ... have any problem with this because the decision lies with the woman ... Screening procedures now by the new law ... are okayed by the government. And you have to ask for a permit – a license by the government whether you can actually offer it to the public or not ... cervical screening, mammography, prenatal screenings by triple test ...

(Obstetrician NL 2)

Here, it should be noted that in July 1996 The Population Screening Act (Ministry of Welfare, Health and Sport 1996) was enacted. This meant that in order to offer serum screening for pregnant women, a physician would have to apply for a permit. This new requirement, asking for a permit and needing a license was viewed as problematic by some practitioners:

To do triple screening you will need a licence ... Now the situation is that the professionals don't want to ask for permission according to that act because they think the permission will be rejected, so they don't even ask ... I don't think that's [good]. I think you should ask permission because now it will stop the research in this field. Because if you don't [ask], you are not allowed to do serum screening.

(Obstetrician NL 6)

This above expert noted that there were approximately 20,000 legal abortions per year that represented 10 per cent of all pregnancies. He also said that there were 300 abortions for genetic reasons. In his view, genetic abortions, although legal, were somehow traumatic for the political parties:

Legal abortions are only allowed in ... an emergency situation ... for the mother [if she] doesn't think [she] can continue the pregnancy. And I think for genetic abortions it's the same case. She says: 'There is ... one reason or another I cannot continue the pregnancy'. But politicians think it's something else. I don't know why, but they don't like talking about it ... They are afraid of it. I think ... how long we have [had] legal abortion ... about 20 years and it has been a trauma for the political parties because they had a lot of discussions about it ...

(Obstetrician NL 6)

In discussing abortion within the context of prenatal screening, some experts spoke about how adopting this practice was seen by some politicians as going down the slippery slope:

> The government has decided that these kinds of issues belong to the domain of individual responsibility because it belongs to the couple involved and in particular, the woman who is pregnant. This is [the] basic thing … So the issue … is … [Is] Down's syndrome grave enough to terminate a pregnancy? It is not a matter of the medical profession … It belongs to [the one] involved. And [there are] ethical implications. If you are … against the termination of pregnancy for whatever reason, you must be against prenatal diagnosis. So our Catholic University (CU) hospital in Nemeigen [which is] under the pope's influence is the latest centre to do prenatal diagnosis … Other arguments grew stronger and stronger and a couple of years ago [when] they (CU) started doing it. And … it was down the slippery slope – you know the eugenics [line] … We know that we test for neural tube defects of chromosomal abnormalities. We know that chromosomal abnormalities are something that mother nature always handles in the form of a spontaneous abortion – in almost all cases except a few … We do not put any pressure on a couple to terminate their pregnancy.
>
> (Obstetrician NL 2)

Another expert claimed that this slippery slope was only in the minds of the politicians:

> Some [ethicists] … are focused on the undesirable effects of offering the option of pregnancy termination as a measure [in] the public health sphere and the problem of the … slippery slope … that people are going to accept fewer and fewer abnormalities in children which is a crazy argument … It simply doesn't exist for people who want to have children. [A] slippery slope around congenital and genetic diseases only exists in the minds of politicians … ethicists and church authorities who are far away from everyday reality of people who want to have children but faced with risk from the family history or previous affected children.
>
> (Medical geneticist NL 4)

A psychologist, who had carried out a study on pregnant women, found that there was a difference between abortion as a woman's right to choose (i.e. not wanting a child) and selective abortion as in prenatal diagnosis:

> I think abortion in the context of prenatal diagnosis is a very different question than abortion if you don't want a child. I asked [women] if there was a difference between these two kinds of abortion and they all said that there was a difference.
>
> (Research psychologist NL 7)

Echoing a similar point, an ethicist claimed:

> I think that [a woman's right to choose] is quietly accepted. People ... realise that [there] are wanted pregnancies and that's very difficult for a woman to end a wanted pregnancy. You should never confront a normal abortion ... [by] I want to get rid of the child [sic] ... I think that people do realise that the message is different in a way because we are talking about wanted pregnancies, which will make the abortion different.
>
> (Ethicist NL 5)

In the above, Dutch experts' claims about reproductive genetics are shaped by the politics of abortion which appear heated at times. While most accepted that pregnant women have a right to choose normal abortions and this is protected in law up to twenty-four weeks, they perceive genetic abortions as a different matter. Genetic abortions are political not normal. This is because in the context of prenatal screening, normal abortions appear as unwanted pregnancies, while genetic ones are perceived as wanted. Here, time reckoning becomes intertwined with legal reckoning as divisions in the social production of reproductive space emerge. On the one hand, if experts press for a screening timetable with pregnant women, experts themselves are co-ordinated temporally and spatially by the law with whom they must reckon. On the other hand, whether women experience normal, legal abortions or genetic political ones, their reproductive time is marked by their own desires to continue or not with their pregnancies. Contrary to their treaters, women's experience of pregnancy compresses reproductive space (i.e. their wombs in their bodies) and reproductive time. Their pregnant bodies and the contents of their wombs are all that there is, as they are marked in a paradoxical space of legal control and personal liberation (i.e. a woman's right to choose).

Creating a genetics moral order: ethics and reflexivity in England

The main overarching cultural theme in England was the need for ethics in reproductive genetics. This was a central thread in the British interviews. One expert, an ethicist herself, believed that the ethical issues in medical genetics are different than in other areas of biomedicine:

> Are ... the ethical issues different in genetics from other areas? I think this might be the case ... One is the attitude to information, because medical ethics [has] traditionally been about giving the patient all the information and more control, whereas genetics has lead to arguments ... such as [the need] not to know ... This is quite a new development. The second area relates to confidentiality and sharing of genetic information between family members ... Genetic information can have such potentially far-reaching implications ... Confidentiality must be strengthened rather than weakened ...
>
> (Ethicist E 9)

Another expert believed that the growth of prenatal technology brings a whole new cultural relationship to ethics:

> The conundrum is ... that thing about ethics and the facts. We are assuming that its mutual – technology and ethics ... It is this technology that brings with it a whole cultural relationship and it's one that science needs to be responsible about.
>
> (Researcher E 1)

For another expert, ethics was linked to a series of issues from choice to commercialisation:

> [Ethics] are to make sure that there is adequate information provided ... One of the aims is to increase choice rather than to terminate more pregnancies ... To allow choice ... is ... an ethical issue. I suppose commercialisation raises the ethical problem ... This captive situation ... raises a big ethical problem ... whether [or not] you call pregnancy a captive situation. In other words [you] introduce the idea of the test to somebody during the

pregnancy ... The idea is to introduce it before there is a pregnancy, if you can. Once the person is there and you say would you
like this, they are likely to say 'Yes' more ... than if they weren't
pregnant at the time.

(Clinical geneticist E 3)

Linked to the issue of choice is the issue of informed consent that
some experts believed was crucial in discussions on ethics. One expert
noted that in the past women were having prenatal screening without
proper informed consent:

[From early research] ... women were not being asked for
informed consent ... They were undergoing prenatal tests without
informed consent. So informed consent is a crucial issue that has
to be addressed right now.

(Researcher E 1)

In this context, another expert claimed that:

Women should be better informed ...

(Public Health Official E 5)

Another expert echoed this concern:

There are women ... that don't know enough about what is actually happening. It is important that they know, that it is a
conscious decision to have the test, [that] they have chance to
think about it ... At each stage of the process, they have a chance
to reflect and consider so at the end of it they can say, yes I did
consider it ... And if the outcome is not exactly what they wanted
... they could say everything was explained to me properly and I
had a reasonable chance to make ... the right decision ... What's
important is, that people have a choice as to whether [or not] to
get into that situation in the first place.

(Epidemiologist E 6)

Nevertheless, another expert thought that informed consent is a
very difficult concept for most people:

We have to deal with ... the concept [of] informed consent,
because ... we have to be very open to patients about [it] ... I think

the whole concept of informed consent is a very difficult one for most people.

(Ethicist E 7)

Linked with this discussion is the idea of how information is communicated in the process of gaining informed consent:

There is a notion that the problem starts with the question: Is the information any good? ... Are the results communicated appropriately? and How much counselling and by whom? and [How much] supervision is required? There are then issues about confidentiality. There is the basic fact that genetic information is familial information and what are the implications of that.

(Ethicist E 8)

Also, another question arose: how is informed consent linked to responsibility towards one's patients?

My philosophy is this that when we have scientific knowledge, when we introduce new things which [were] not possible before ... it creates responsibility ... From the time that that's preventable within the medical and scientific systems, somewhere there is a responsibility for every movement ... This responsibility should be constant to the patients because if it's not, if they have not been informed that they are at risk and what their options are [then] they have an affected child without being informed. That is absolutely catastrophic and in this country they all tend to sue the ... doctor, the obstetrician.

(Clinical geneticist E 4)

Other experts link ethics with the public understanding of science. Here, one believed that scientific facts should not be separated from ethics:

When the public becomes well informed about scientific facts, then we will be in a better position to make ethical and moral judgements about many problems ... There are certainly a lot of people and writing and questioning ... but [ethics) is not the dominant issue in the public arena. We are not an informed public.

(Researcher E 1)

Other experts echoed their concern about an uninformed public:

> The public is largely informed by what they read in the papers, which is misleading ...
>
> (Ethicist E 9)

and

> [We need] to increase the level of public awareness and their knowledge ... It means to have anything to do with genetics ventilated in the widest possible way ... making knowledge freely available.
>
> (Clinical geneticist E 3)

Another ethical issue was the failure to deliver equitable services:

> We have failed to be interested in the real ethical problem, which is our failure to deliver the service. So it is delivered inadequately and inequitably. That's my ethical perception.
>
> (Clinical geneticist E 4)

In England, concerns for ethics shape experts' claims about reproductive genetics. However, some claim that genetic ethics are different from bioethics. This is because conventional bioethics enforces the need to know while genetic ethics may uphold the need not to know. Also, bioethics defends sharing of information. Genetic ethics does not because it considers the far-reaching temporal and spatial implications of genetic knowledge. Experts perceive genetics ethics as facilitating the acceptance of technologies in the public domain as well as establishing a normative framework for the institution of reproduction. Ethics assist the construction of a genetics moral order in which pregnant women's bodies become viewed as reproductive, albeit captive resources. Reproductive genetics thrives on the moral mediation as well as scientific intervention of biomedics. British experts claim that they need to exert responsibility. But, more importantly, public accountability and informed consent are needed to ensure the circulation of a genetic moral order. In this moral order, reproductive genetics may inevitably cause disruptions and transformations in the temporal links between pregnant women. Some may go to term, others may not. Nevertheless, experts implied that pregnant women need time to make the right (i.e. ethical) decision before these temporal links are broken. Thus, for pregnant women, reflexivity or making time to make ethical decisions emerges as an important piece of reproductive time.

Setting reproductive clocks for maternity care in Finland

The overarching cultural theme was care for pregnant women. In Finland, a maternity care system has been in existence since the 1940s and experts claim that this system provides an appropriate infrastructure for prenatal screening and diagnosis, offered to a captive treatment population – pregnant women. Experts are unanimous that this system is very well received and appreciated. One expert said that this system has been the backbone of developments in primary health care, while another expert spoke of the trust that pregnant women had and the high uptake amongst this group:

> When people [sic] get pregnant, they get their first visit at maternity health [centres]. Ninety-nine per cent of women go there before 15 weeks ... because ... they get money and ... all the care that they [need] ... They rely very much on these centres; they trust them ... They go there ... those who are pregnant for the first time ... at once ... They will [go] when [they] are ... 10 or 12 weeks ... That is the common practice now ... For the first visit ... they will get to know about these things [i.e. prenatal diagnosis]. If the [woman] is over ... 40 years, the health care nurse will tell them that there is a possibility of having amniocentesis because of [the] chromosome risk ... If there is serum screening, they will tell [them] about it ... They are very conscious about these things ... They pass [on] the knowledge because it's their job.
>
> (Medical geneticist F 16)

Another expert implied that trust meant that the system should be cautious:

> If something is offered ... especially for pregnant mothers or for small children in the [maternity] ... care system, the great ... majority will accept it ... In that sense, [the] Finnish health care system should be very cautious in what it offers because if it offers something, everyone takes it ... I think the trust in the ... system is extremely strong ... If this system offers something, it means that it is ... good. It won't offer anything bad ... I think [generally] our

> health care has not had any ... big catastrophes ... people really do trust [it].
>
> (Medical geneticist F 5)

This view is consistent with the claim that Finns generally trust health professionals:

> There has never been a big public outcry on any of these [screening] issues. Usually Finns – again this is a very different country from most of Europe – trust the health professionals who are offering a service. For instance, immunisation – there is nothing statutory ... it is all voluntary, but people follow [it] 100 per cent. They trust the system ... There is tremendous trust in the work of public health nurses who carry out these practical things.
>
> (Policy maker F 13)

One expert claimed that when women come forward for maternity care, they are urged on by their need to have healthy babies:

> [Pregnant] women ... want to have healthy children ... They want to have any test that will tell them that their baby is OK. This is a basic, endless need. They will have as many tests as possible, if they can ... if somebody else pays. I'm sure ...
>
> (Medical geneticist F 4)

This free maternity system has provided an infrastructure for prenatal diagnosis and screening:

> Women know the medical [system] ... Because of the system which started in the 1940s ... they ... unite the screening system to that [system] ... [For them] this is a natural system ... They don't know that [this system] is put into the law [or] think ... it as a political matter. They have been served before without screening and when the screening started, they noticed only that it is extra to the old system. I think that is the way ... they ... see it.
>
> (Policy maker F 6)

and

> 100 per cent [of pregnant women] will get the information [about screening] when they go to the maternity health care centre ...
>
> (Medical geneticist F 16)

and

> There is ... [a] kind of consumerism here ... because the network
> is there ... You ... have a basic program ... It is very difficult to say
> ... 'Oh thank you, I don't want [the screening]' or 'I [won't] take
> this or ... that' ... The more you offer things, the more people
> consume them. So if you are screening ... or if you offer these
> services, people will use them.
>
> (Ethicist F 15)

But one expert said the 'carefree days [i.e. early days of the maternity system] are gone'. While women are inducted in a system in which
prenatal screening has really made it, they are not prepared if problems arise:

> The care [free] days are gone ... It's not only chromosome analysis
> but ultrasound that really has made it ... Almost everyone [preg-
> nant women] will have ultrasound and it will reveal quite a lot of
> things ... and [sometimes] very unexpected [things] ... Because very
> often people [sic] when they ... [hear about] ultrasound, they will
> go [to] see the baby. They [are not] really ... prepared ... that ...
> problems might be found.
>
> (Medical geneticist F 16)

In a similar context, another expert says:

> Traditionally we have had ... rhesus screening. For over 10 years,
> we have had Down's screening offered to mothers over 35 or 37 –
> depending on the place ... One thing which is well established and
> offered to all pregnant women [is] ultrasound screening. This has
> other purposes [and] ... sometimes ... one finds ... genetic malfor-
> mations ... A rather new thing is [Maternal Serum Screening] for
> Down's syndrome.
>
> (Medical geneticist F 5)

Another expert claims that more thought needs to be given to the
issue of population screening:

> We need to think very carefully about whether we want popu-
> lation based screening ... whether we want it [where] we want to
> put it ... We ... have certain units mainly for genetics. Should we

keep population screening ... for families with a defined genetic problem?

(Medical geneticist F 16)

As seen from the above, the existence of the Finnish maternity care system since the 1940s shaped experts' claims about reproductive genetics. In the eyes of experts, pregnant women are marked as much by their need to consume as their need to reproduce. These consumption needs emerge from their trust in the long established, well-serviced maternity care system, while their reproductive needs revolve around reproducing healthy children. What is of interest is that few experts mentioned problems and when these arose, they were contextualised in the context of maternal age. Thus, it could be said that for Finnish women, reproductive time is employed to best advantage when young rather than old bodies reproduce. Temporal formulations for normal reproduction are organised at a collective level. Pregnant bodies are synchronised and harmonised into a state wide, servicing system in which age becomes an anchor. Rather than a source of energy, time becomes somewhat of a burden for those whose biological clocks are perceived as not synchronised with their reproductive ones.

Making expert claims and marking reproductive time

The findings from the experts' study reveal overarching cultural themes, existing in each country. In earlier work (Petrogiannis *et al.* 2001) on a review of policy, law and ethics in the four countries, differences as well as similarities in respect of practices and regulations with regards prenatal screening had already emerged.

In Greece, thalassaemia screening has been viewed as a medical achievement because the total number of children born with thalassaemia has diminished significantly. As a result of this triumph, Greek people accept other forms of prenatal screening wholeheartedly. The thalassaemia success story brings into focus the medical profession's achievements with prenatal technologies whether in urban or rural contexts. Reproductive time is employed to best advantage when the genes of pre-nuptial couples rather than the bodies of pregnant women are the spotlight of concern. Although central to reproduction, pregnant bodies become non-places in reproductive genetics.

In the Netherlands, the Population Screening Act (Ministry of Welfare, Health and Sport 1996) reveals that there is resistance to prenatal screening on a wide scale. This resistance is related to the Dutch approach to childbirth and delivery and the wide acceptance of home deliveries performed by midwives. Here, the politics of abortion shape experts' concerns, as they make a distinction between normal and genetic abortions. There is a division in the social production of reproductive time. On the one hand, time reckoning for experts becomes linked with legal reckoning. On the other hand, time reckoning for pregnant women is linked with their own desires. Caught in a contradiction between the law and their right to choose, Dutch women may find that reproductive space and time becomes compressed.

In England, similar to the Finnish situation, prenatal screening is organised and financed locally and much of the emphasis in the public debates is on pregnant women's right to choose (Petrogiannis *et al.* 2001). For English experts, genetic ethics, differed from bioethics. Genetic ethics helped to facilitate the advancement of new technologies and a genetics moral order. In this social space, experts needed to exert responsibility, while pregnant women needed reflexive time to make the right (i.e. ethical) decision.

In Finland, prenatal screening has proliferated through municipal maternity care centres that are attended by 99 per cent of pregnant women (Petrogiannis *et al.* 2001). Finnish people tend to accept most services that are on offer by the Finnish health care system. This explains why such a large percentage of pregnant women attend local maternity clinics. Finnish maternity care shaped experts' claims about reproductive genetics. As consumers and reproducers, pregnant women's needs were marked by their trust in the maternity care system and their desire to reproduce healthy children. Reproductive time was employed to best advantage when young rather than old bodies reproduced. This was because in Finland temporal formulations for normal reproduction were organised at a collective level.

In conclusion, in the transition from hospital to surveillance medicine, culture plays an important role in establishing the visibility of specific illnesses and in shaping the time and spaces for reproduction in particular societies. While time is important in the social construction of subjectivity (Zerubavel 1982, 1997), marking reproductive time in the context of the above four European societies, constructs social relations within the institution of reproduction, albeit how reproductive time is marked or standardised varied between countries. While all bodies are forward orientated in relation to survival and meaning (Shilling 1993: 179), pregnant bodies are not only forward orientated but also synchronised in the present by reproductive time, co-ordinated by technological interventions. Indeed, prenatal technologies reflect huge disruptions and transformation in the 'temporal' links between pregnant bodies and foetal bodies.

Nevertheless in all countries, it is clear that reproductive time, similar to time itself (Adams 1998) is becoming increasingly commodified and marking reproductive time involves complex gendered practices. Simply, reproductive time is becoming product oriented. It is progressively more linked to outcome (i.e. the production of normal, non-disabled babies) and the bodies of women, not men, are temporally and spatially co-ordinated in this surveillance process. Here, I ask, what else is new? I have no answer. Nevertheless, I contend that an understanding of how experts as key players construct reproductive bodies and the sorts of strategies they use to mark reproductive time are important issues. This awareness helps us to further understand current cultural and clinical practices being developed in reproductive genetics. The more we uncover the somewhat invisible practices of surveillance medicine, the more we challenge the inequalities that become embedded in the institution of reproduction. Let us use time as a resource, as we begin to challenge experts' marking of reproductive time.

7 Reproductive genetics and the need for embodied ethics

> The activities and theorising of the scientific community do not proceed in a vacuum. They are subject to all the biases current in the established social system; these affect, and sometimes warp, their conclusions.
>
> Evelyn Reed, *Sexism and Science*, p. 7

Introduction

In order to illustrate how reproductive genetics works through pregnant bodies by medical surveillance, technologisation and control, I have examined in this book the claims of experts as genetic storytellers – active knowledge builders and the subtle power of their knowledge over female bodies. As key actors in the reproductive genetics arena, experts aim to improve the results of the reproductive processes through their knowledge of genetics, prenatal technologies, potential disease, foetal abnormalities and heredity. Their biomedical knowledge becomes embedded in prenatal genetic technologies and is played out on the bodies of pregnant women.

Regardless of their primary medical discipline or practice (i.e. as obstetricians, policy makers, medical geneticists, etc.), these experts are partial – not impartial – genetic storytellers. Their knowledge and its technological application is constructed within the particular constraints of an expert's discipline; an expert's work culture; the particular stakes that the expert has in producing this knowledge; and the internal criteria of science (pragmatic, utilitarian, etc.) that the particular expert uses.

Tracing the development of the notion, expert Zygmunt Bauman (1987) contends that the emergence of modernity was a process of transformation from wild cultures into garden cultures. This latter 'cultivated' culture, rooted in rationality, initiated a new structure of domination: the rule of knowledge and knowledge as a ruling force. This culture (i.e. modern society) characterised self-conscious culture – a culture in need of 'know how'. Garden culture could only be sustained by specialised personnel whose roles were to be designers, supervisors, 'surveillors' and most importantly, educators. Within garden culture, the notion of the expert emerged. This social actor, the expert, not only became the conduit for a new kind of power, both pastoral and proselytising, but also was armed with exceptional skills. The expert was the specialist in bringing human beings up to the level of perfection required by the common good.

As experts in reproductive genetics articulate, construct and reproduce their positions of authority, they fulfil important social and scientific roles as interpreters of genetic knowledge. Experts represent specialists: reproductive invigilators, fulfilling the cultural need in the population for genetic supervision and 'know how'. In this invigilation process, they embody educators, surveillors and storytellers whose role is to reinforce and legitimate a genetic moral order.

In this final chapter, I will first offer some reflections on my experiences as a feminist and sociologist, doing the experts' study. Then arguing that experts' ethics are disembodied ethics, I will review their differing ways of seeing ethics and contrast their views to embodied ethics. Lastly, I argue that we need to revision women's experiences within reproductive genetics.

Reflections on the experts' study

When I first started to do the interviews, one of my colleagues said to me: 'Elizabeth, how can you carry out these interviews when as a feminist, you don't agree with a lot of the work these experts do?' At the time, I thought the question was an ignorant one, and I still do after having carried out the interviews and done the work. I believed very strongly throughout the study that dialogue was possible even if, as a feminist sociologist, I disagreed with an expert. More importantly, my research role was not to be a preaching feminist but to be an open-minded, listening interviewer. I wanted to gain valuable information about experts' reproductive genetics work. I could not perhaps see through their eyes, but at least if I listened to what they were saying, I could begin to understand the sorts of values they upheld in their practices, their views towards pregnant women and on the workings of reproductive genetics.

If I started disagreeing passionately with an interviewee during an interview, gaining valuable information would be difficult indeed. I did not disagree verbally with the expert interviewees, although sometimes

I had to bite my tongue when listening. This does not mean that after the interview, I did not disagree with an expert. This happened on a few occasions. While my role was a listening one, I found it sometimes easy and sometimes difficult to listen about work in a field that I was becoming more familiar with. It was difficult to listen sometimes because I had the distinct feeling that the interviewee was totally absorbed with himself/herself. That always troubled me because I could not understand how the particular expert would be able to listen to patients – pregnant women, couples or families. But perhaps, this absorption with themselves allowed them to focus totally on their work and excel in their area. Sometimes, I felt an expert (usually a male expert) had rather cavalier attitudes towards women and that saddened me. I wondered how his women patients would cope. As I listened to experts speak, I gradually heard that they had their own words and their own truths.

Experts ranged from those who were very enthusiastic about the use of genetics in reproduction to those who were very sceptical. It was also interesting how some experts asked me who else I was going to interview in their country. They wanted to know if I was interviewing any of their close colleagues or friends. I did not tell them on the grounds of confidentiality, but others found out because of their professional contacts. One of the most interesting research experiences for me was the fact that these expert respondents held what in my view were conflicting views.

For example, I was openly surprised when one expert talked about how prenatal genetic technologies could have a devastating effect on families psychologically. Although, she herself was a medical doctor and now is a government minister, she pointed a finger at 'medical doctors' and in particular obstetricians to find 'a balanced way':

> We have to find the moderate, a balanced way ... to all of these problems, prenatal screening is the most important ... because it costs and because it is a disaster for the family psychologically and also a disaster for the state and the social security system.
>
> (Policy maker G 10)

After the interview, I discovered that she had a relative who had a terrible experience with amniocentesis and as a result went abroad to have her baby. On the other hand, she was not keen on disabled babies:

> [think of] handicapped people ... if they are treated ... there is a responsibility [here] ...
>
> (G 10)

In this context, this expert said she was not happy about disabled babies being born because her country is not friendly to disabled people.

This sort of thing happened many times. An expert would reveal compassionate views and then in the next sentence say something that jolted me or in my view, seemed a contradiction to their compassionate views. I remember one expert who was a leading medical geneticist in his country talk sympathetically about his patients and then in the next sentence talk about how he disliked disability. But perhaps, this was not a contradiction for them. They were medics and wanted to eliminate disease – aborting potentially disabled or genetically diseased babies was part of this mindset.

Disembodied ethics versus embodied ethics

It is interesting to note that experts' views tended to be clear and articulate especially when they spoke about ethics. Besides embodying educators or surveillors experts, as mentioned in Chapter 2, are meant to embody ethics. But, ironically, the sort of ethics they are meant to embody is a highly rational disembodied one. In most instances, they give too little attention to personal relationships and embodied experiences. Many experts were keen to share their views on ethics, and as one expert said:

> We have very little discussion of ethics in medical school. We learn it when we decide on difficult problems in our work.
>
> (Obstetrician NL 3)

It became very clear to me in my interviews with experts that they held many opinions on how ethics worked in their practices and these views tended to be shaped by their differing societies, areas of work and cultural priorities. The significance and role of ethics varied for them (see Chapter 2.) Experts viewed ethics as an essential to their rational way of thinking; as relative, meaning that ethics were conditional or changeable from society to society and/or as problematic, given that within the developing area of reproductive genetics new sorts of moral justifications emerged. Ethics became for them an area that needed to be explored, but I found that they tended to be conservative in their explorations. Ethics tended to be used to protect their positions in clinical settings rather than those who they treat.

The importance of ethics

The majority of experts believed that ethics were important in their work. Many said that they had to keep ethical issues in the forefront of their minds. When they talked about the significance of ethics, they mentioned a variety of ethical aspects in their work including: the voluntary nature of tests, sex selection, the importance of giving information to patients, the correct justification to screen and the ethics of population screening.

Experts who emphasised the importance of genetics tests or prenatal screening as being voluntary implied that there should be no form of compulsion. They stressed that patients' decisions should be voluntary and experts emphasised informed choice.

> The main ethical thing ... is that all these [genetic] tests have to be voluntary ... We should not take any stand [as to] whether [or not] someone should have it or not.
>
> (Clinical geneticist F 16)

and

> I think one of the considerations that we need to take into account ... it would be voluntary in any case ...
>
> (Policy maker F 13)

The following paediatrician's views were shaped within the context of thalassaemia in Greece. He recognised that there are 'some ethics' because of the problem of reducing the number of children with thalassaemia. He suggests that ethics meant for him recognising the importance of patients' making voluntary decisions and physicians not influencing these decisions:

> We have to do the diagnosis ... in order to have prevention. It is not real prevention actually. But it is limitation [or] reduction of the number of affected children. If you want to prevent or reduce the number of those with ... thalassaemia, we have this procedure. There are some ethics because this is the problem [reduction of the number of affected children] and what we discuss with the parents ... We have to discuss with the parents and they have to take their own decision. We don't want to influence ... We say that's what we can offer.
>
> (G 6)

In this context of making voluntary decisions, another expert, a clinical geneticist, believed that providing enough information was ethically important so that patients 'know beforehand' in order to make informed and voluntary choices.

> We should be ethical enough to explain carefully everything and have the time to explain everything so that the parents have a chance of deciding about things, because it's impossible to decide about something that you don't really know about ... So this is one ethical thing, because we think that you should know before-hand that there is going to be the test results and if you really want to take this test.
>
> (F 16)

But another expert indicated that problems could arise with this sort of information because there was so much of it – 'a stifling amount' – in order to choose. This clinical geneticist said:

> Now when you look at the amount of different types of informa-tion that is used in our screening programme it is formidable because you've got different methods and to sensitise people and to make them aware [about] screening ... that there is a choice ... It is a stifling amount of information ...
>
> (E 4)

I found that most experts were not supportive of sex selection only for certain genetic disorders. In this context, Wertz and Fletcher (1998, 1993) argued that prenatal sex selection was an ethically and socially troubling problem and they found experts who were open to its use over and above extraordinary cases. On a global basis over a nine-year period, they found a greater willingness amongst experts to perform or refer a patient to sex selection. One medical geneticist talked to me about sex selection because she had just seen parents with haemophilia in their family. She felt that there are many ethical issues involved in prenatal diagnosis and implies that, in exceptional cases for certain genetic disorders, it would be necessary to reveal the sex of the foetus.

> There are ethical issues [where] ... you have the facility or the tech-nical [know how] to find something and it depends let's say about the foetus where it is a sex linked disorder. A few minutes ago, it was a haemophilia case. It is of course important to know about the foetus – to go on if it's a boy to prenatal diagnosis is not normal

but abnormal because you can very easily find the sex of the
foetus. I don't believe that this is good practice or makes sense
[generally]. There are many ethical issues in prenatal diagnosis ...

(G 5)

I remember also talking to another expert, an ethicist who reflected
upon her experiences on a local ethics committee. She talked about the
difficulties with sex selection. She also points out that while this practice may be unethical, it is not illegal. She talks about the acceptable
side of genetics and ends on a cautionary note because 'we don't know
what the effect of those tiny changes may be':

> We were talking about termination of pregnancy and we were
> reviewing ethical issues on things like sex selection and the termi-
> nation of pregnancies that are not acceptable. That is interesting
> because we feel that is unethical but the interesting thing is that it's
> not illegal. The ... law is written in such a way that ... we have
> abortion on demand ... The risk to the mother of continuing with
> the pregnancy is higher than the risk of just termination. That
> means that you can ... almost always find justifications ... If her
> reason is that it's a girl but she wants a boy, it's arguably legal. But,
> if it was genetic, then I will argue that sex selection is probably
> unethical except where there is [genetic] risk ... I don't believe that
> is necessarily wrong. I think that's what most of us will regard as
> the acceptable side of genetics. But, we also have to recognise what
> we don't know we're doing – we don't know what the effect of
> even those tiny changes in the broader genetic terms there may be
> on the planet of human kind generally and those are important
> issues.

(E 7)

Some experts with the view that ethics was essential in their work
also felt it was important to have the correct justification to screen, but
this was a complex issue. One gynaecologist illustrated this point. He
felt that regardless of whether or not Down's syndrome was severe
enough to screen for, 'what the parents want' was 'the only thing you
find important':

> They always concentrate on the point ... 'Is it worthwhile?' Is the
> disability you screen for; is it worthwhile to screen for? Is it? Who
> gives you the right to do that? Is Down's syndrome ... severe
> enough to screen for? We don't know and we always point out ...

the only thing that you find is important is what do the parents want ... But, the other side of it is, 'Did you ask the handicapped people?' ... No, we didn't. And I think that that's the wrong person to ask because you are asking 'Do you want to be born?' And no one can say no, I don't like it. But that's the main point in ethical issues. Is it worthwhile or is it justified to screen?

(NL 6)

On another level, another expert, a gynaecologist, appeared to be acutely aware that if all pregnant women were given maternal serum screening, this type of mass screening presents ethical problems. He was echoing some of the views in his country, the Netherlands:

As soon as the mass aspect [of prenatal screening] comes in ... It's not a financial thing. It is an ethical thing ... The thing is that many arguments are put against it. One of them is that if you strictly apply the requirements that a screening technique has to fulfil – then the triple test does not come up to that standard.

(NL 2)

Ethics as relative or problematic

Experts believed that ethics, while somewhat important, were also relative. At times, an air of relativism pervaded their ethical terrain. This type of relativism is confirmed by the following obstetrician when he said:

What is ethical to me maybe is not ethical for you ... What is ethical in Greece to terminate the pregnancy with [Down's] may be unethical in another country. It's very interesting point that I think that we can not make rules for everything. Is it ethical to do a foetal reduction – to reduce the number of the foetuses from 5 to 3? And why is unethical to reduce in these kind of cases? It is not unethical to terminate a pregnancy because the law here says that everybody can do a termination up to 12 weeks. What does up to 12 weeks mean? So it is ethical ... to terminate a pregnancy because the woman doesn't want it?

(G 3)

In this context, another expert felt that ethics were on the one hand already known by him and his colleagues, while being on the other hand, unresolved. Above all, there was no need to discuss them. There was no resolution:

> Most of the meetings I go to, they don't [discuss ethical issues] … Probably it should be more explicitly discussed. I think they're not [discussed] because there is a sense that there is actually no resolutions to them and therefore it's [like], they're kind of, already known. There is no resolution and it's therefore tedious to go over them again, I do feel it's that sort of [thing].
>
> (Epidemiologist E 6)

Still, one expert questioned whether or not experts could 'hold a view' on ethics, as one expert asked:

> Are there any expert views on ethics apart from those very technical philosophers with fairly technical philosophical questions?
>
> (Policy maker E 8)

Emphasising this problematic area of ethics *vis-à-vis* genetics, another expert, a general practitioner, was totally perplexed by new developments. He asks: 'What will be the limit?'

> The possibilities to look at diseases [are] becoming very big. There are many diseases you can theoretically look after when you are pregnant. But, it is an ethical problem – what will be the limit?
>
> (NL 1)

The above experts framed their ethical concerns in highly rational disembodied contexts. While these concerns are of interest, they tended to be somewhat dispassionate. Experts wanted screening and tests to be voluntary; they questioned sex selection in populations; they saw the importance of accurate information and they believed that there is a correct justification to screen. Whether or not they saw ethics as problematic or relative, some questioned the development of even more practices in reproductive genetics and having an expert view on ethics. Most strikingly their ethics were about deciding, thinking, considering or justifying. An awareness of corporeality was lacking.

I contended earlier in this book that there is a need in reproductive genetics for responsible ethics framed by and through embodied relationships. I meant that when experts deal with pregnant bodies, they should be aware not only that these bodies are gendered but also that

the corporeal changes these pregnant bodies experience have implications for their gender as well as genetic risk identity whether they
become good or bad reproducers.

Embodied ethics take as its starting point the corporeal experiences
of moral, gendered individuals and with regards reproductive genetics,
this starting point is crucial if women's bodies are to become whole not
broken. Given the above, an embodied ethics is absent from this field.
Consistency (Kuhse and Singer 1999) and factual accuracy not embodiment and emotion are the requirements of defensible ethical positions.
In the field of reproductive ethics, moral analysis aims to bring agree-
ment by the use of rational argument (Bayles 1984: 3). Emotions are
not part of the equipment to discern moral answers (Little 1996).

A feminist approach to ethics and specifically bioethics challenges
this rationalised view of morality. A feminist approach to bioethics
evaluates medical practices in terms of the impact of such practices on
women (Lebacqz 1991) and their bodies, helps experts to recognise
that gender matters and insists that women's varied experiences should
be taken into account (Rothenberg 1996) – experiences, I would add,
that are corporeal. Embodiment must be a key issue in feminist
approaches to bioethics. Here, similar to Anne Witz (2000), who refers
to Kathy Davis's (1997) work, I have been more concerned in this book
with how women's bodies (in this case pregnant bodies) are controlled
and broken through patriarchal practices rather than problematising
embodiment *per se*. Thus, I would contend that embodied ethics with
special reference to reproductive genetics makes moral appraisals of
the impact of a variety of medical practices and technologies on pregnant bodies. Simply, women's bodies and their corporeal experiences
are the starting point for any ethical evaluations; these gendered bodies
and experiences are not taken for granted.

A way forward: revisioning embodied, gendered experiences

While the book's concentration has been on experts and their claims
about reproductive genetics, the analytical focus has been more on the
workings of gender inequalities and in part, disabilism and less on the
workings of other social inequalities. Of course, when pregnant
women encounter reproductive genetics, their experiences are shaped
by all the social inequalities that embed them in society such as class,
race, ethnicity, sexuality and so on. I have maintained that to be truly
sensitive to all pregnant bodies, experts working within reproductive
genetics should develop a focused gender dimension in their work.

They may uphold the view that prenatal genetic technologies are gender neutral and not gender biased. But, this view should be challenged, given that technologies affect our conceptions of femininity and masculinity (Wajcman 1991; Stabile 1994), technologies are feminised and masculinised as they take shape and technologies are shaped by gender (Rudinow Saetnan 1996). Furthermore, the sorts of technologies used in reproduction are designed for women's bodies and have profound consequences for gender relations during pregnancy and beyond (Faulkner 2001).

The injection of biology into social relationships through reproductive genetics allows more attention than every before on the workings of the 'female' human body in reproduction (Franklin 1993) as well as this gendered body, situated in other areas of social relationships (Martin 1992). I have viewed these developing discourses on the interplay between nature and human biology as raising important issues around the discourse on the body. At the intersection of surveillance medicine and reproductive genetics is a focus on the female body, interacting within the community, the site of genetic capital and a material entity where scientists are able to 'see' the structure of the material of genes – DNA – as well as the growth of the foetus. But, this physical body, the original site for foetal (Newman 1996) and/or genetic investigations, is shaped by gender. It is a pregnant body – a gendered, female body. Through the medical gaze, this body becomes less than body, as it is 'relegated to foetal environment' (Degener 1990: 80).

That a pregnant woman's genetic capital may be ranked in this gender-biased context suggests that more female than male bodies are subject to a reproductive morality imposed by experts. As we have seen, the imposition of a genetics moral order during reproduction has the effect that pregnant bodies are separated into good and bad reproducers, that some pregnant bodies experience their reproductive potential as shameful in comparison to other pregnant women and that a body reproducing badly can be seen as a new form of embodied transgression for women. The bad reproducer has 'bad genes' and dares to allow 'bad genes' or 'gene mistakes' come into this world.

In this book, I have attempted to revision women's experience of reproductive genetics. Revisioning means letting go of how we have seen in order to construct new perceptions (Clarke and Olsen 1999). Thus, we need to let go of our damaging images and ideas about gendered bodies as bad reproducers in order to construct new perceptions about them and their embodied experiences within reproductive genetics.

In her work as a feminist philosopher, Marilyn Frye (1983: 76) defines 'the loving eye' in contrast to 'the arrogant eye':

> The loving eye is one that pays ... attention ... This attention can require a discipline ... one of self-knowledge, knowledge of the scope and boundary of the self ... It knows the complexity of the other as something, which will forever present new things to be known ... The loving eye seems generous to its object...for not being invaded, not being coerced, not being annexed must be felt in a world such as ours as a great gift.

Too often, the bodies of pregnant women have been broken by 'the arrogant eye'. While these women have been helped, they have also been harmed. In this book, I have attempted to inject new ideas on gender and the body into an analysis of reproductive genetics. I have argued that our Western ethics have excluded female bodies from full moral agency during pregnancy and this exclusion has resulted in a fragmented morality of the female body. Simply, in moral terms, women's bodies are not whole; they have become 'broken'. In bearing witness to pregnant bodies who experience reproductive genetics, let us use 'the loving eye' and help to make their bodies whole.

As more prenatal technologies are being deployed, critical scholars need to place themselves at their 'foetal work stations' (Haraway 1997: 35), make visible the multiple sites of contestation in the new genetics and expose some of the repressive dynamics of reproductive genetics.

Bibliography

Abberley, P. (1996) 'Work, utopia and impairment', in L. Barton (ed.), *Disability and Society: Emerging Issues and Insights*, Harlow: Longman.

Adam, B. (1998) *Timescapes of Modernity: The Environment and Invisible Hazards*, London: Routledge.

Advisory Committee on Genetic Testing (2000) *Prenatal Genetic Testing*, London: Department of Health.

Al Mufti, R., Hambley, H., Farzaneh, F. and Nicolaides, K.H. (1999) 'Investigation of maternal blood enriched for fetal cells: role in screening and diagnosis of fetal trisomies', *American Journal of Medical Genetics* 85: 66–75.

Andrews, L. and Nelkin, D. (1998) 'Whose body is it anyways? Disputes over body tissue in a biotechnology age', *The Lancet* 351: 53–7.

Anker, S. (2000) 'Gene culture: molecular metaphor in visual art', *Leonardo* 33, 5: 371–75.

Annandale, E. and Clark, J. (1996) 'What is gender? Feminist theory and the sociology of human reproduction', *Sociology of Health and Illness*, 18, 1: 17–44.

Arditti, R. Brennan, P. and Cavrak, S. (1980) *Science and Liberation*, Montreal, Black Rose Books.

Arditti, R, Klein, R.D. and Minden, S. (eds) (1984) *Test-Tube Women: What Future for Motherhood?*, London, Pandora Press.

Armstrong, D. (1987) 'Bodies of knowledge: Foucault and the problem of human anatomy', in G. Scambler (ed.), *Sociological Theory and Medical Sociology*, London, Tavistock.

—— (1995) 'The rise of surveillance medicine', *Sociology of Health and Illness* 17, 3: 393–404.

Asch, A. and Geller, G. (1996) 'Feminism, bioethics and genetics', in S. Wolfe (ed.), *Feminism and Bioethics: Beyond Reproduction*, New York and Oxford: Oxford University Press.

Asch, D.A., Hershey, J.C., Pauly, M.V., Patton, J.P., Jedrziewski, M.K. and Mennuti, M. T. (1996) 'Genetic screening for reproductive planning: methodological and conceptual issues in policy analysis', *American Journal of Public Health* 86, 5: 684–90.

Atkin, K. and Ahmad, W. (1998) 'Genetic screening and hemoglobinopathies: ethics, politics and practice, *Social Science and Medicine* 46, 3: 445–58.

Bailey, L. (2001) 'Gender shows: first time mothers and embodied selves', *Gender and Society* 15, 1: 110–29.

Bailey, R. (1996) 'Prenatal testing and the prevention of impairment', in J. Morris (ed.), *Encounters with Strangers: Feminism and Disability*, London: The Women's Press.

Balsamo, A. (1999) 'Public pregnancies and cultural narratives of surveillance', in A.E. Clarke and V.L. Olsen (eds), *Revisioning Women, Health and Healing: Feminist, Cultural and Technoscience Perspectives*, New York and London: Routledge.

Barnes, C. (1992) 'Institutional discrimination against disabled people and the campaign for anti-discrimination legislation', *Critical Social Policy* 12, 1: 5–22.

Barnes, C. and Mercer, G. (eds) (1996) *Exploring the Divide, Illness and Disability*, Leeds: The Disability Press.

Bartky, S.L. (1990) *Femininity and Domination*, New York: Routledge.

Barton, L. (eds) (1996) *Disability and Society: Emerging Issues and Insights*, London: Longman.

Bauman, Z. (1987) *Legislators and Interpreters*, Cambridge: Polity Press.

—— (1993) *PostModern Ethics*, Oxford: Blackwell.

Bayles, M.D. (1984) *Reproductive Ethics*, Englewood Cliffs, NJ: Prentice-Hall.

Beauchamp, T.L. and Childress, J.F. (1994) *Principles of Biomedical Ethics*, New York and Oxford: Oxford University Press.

Beaulieu, A. and Lippman, A. (1995) '"Everything you need to know": How women's magazines structure prenatal diagnosis for women over 35', *Women and Health* 23, 3: 59–74.

Beazoglou, T., Heffley, D., Kyriopoulos, J., Vintailoes, A and Benn, P. (1998) 'Economic evaluation of prenatal screening for Down syndrome in the USA', *Prenatal Diagnosis* 18, 12: 1241–52.

Beck, U. (1992) *The Risk Society: Towards a New Modernity*, London: Sage.

Beech, R. (1995) 'Using operational research modelling to improve the provision of health services: the case of DNA technology', *International Journal of Epidemiology* 24, 3: S90–S95.

Beekhuis, J.R. (1993) *Maternal Serum Screening for Fetal Down's Syndrome and Neural Tube Defects: A Prospective Study, Performed in the North of the Netherlands*, Groningen: National University.

Benhabib, S. (1990) 'Epistemologies of postmodernism: a rejoinder to Jean Francois Lyotard', in L.J. Nicholson (ed.), *Feminism/Postmodernism*, London and New York: Routledge.

Birke, L., Himmelweit, S. and Vines, G. (1992) 'Detecting genetic diseases: prenatal screening and its problems', in G. Kirup and L. Smith Keller (eds), *Inventing Women: Science, Technology and Gender*, Cambridge: Polity Press in association with Open University Press.

Blane, D., Brunner, E. and Wilkinson, R. (1996) 'The evolution of public health policy: an anglocentric view of the last fifty years', in D. Blane, E.

Brunner and R. Wilkinson (eds), *Health and Social Organization: Towards a Health Policy for the Twenty-First Century*, London and New York: Routledge.

Bobinski, M.A. (1996) 'Genetics and reproductive decision making', in T. Murray, M.A. Rothstein and R.F. Murray (eds), *The Human Genome Project and the Future of Health Care*, Bloomington, IN: Indiana University Press.

Bordo, S. (1993a) *Unbearable Weight: Feminism, Western Culture and the Body*, Berkeley, CA: University of California Press.

—— (1993b) 'Feminism, Foucault and the politics of the body', in C. Ramazanoglu (ed.), *Up Against Foucault: Explorations of Some Tensions Between Foucault and Feminism*, London and New York: Routledge.

Braidotti, R. (1994) *Nomadic Subjects: Embodiment and Sexual Difference in Contemporary Feminist Theory*, New York: Columbia University Press.

Breslau, N. (1987) 'Abortion of defective fetuses: attitudes of mothers of congenitially impaired children', *Journal of Marriage and the Family* 49 (November): 839–45.

Bricher, G. (1999), 'A voice for people with disabilities in the prenatal screening debate', *Nursing Inquiry* 6: 65–7.

Brook, B. (1999) *Feminist Perspectives on the Body*, London and New York: Longman.

Browning-Cole, E. and Coultrap-McQuin, S. (eds) (1992) *Explorations in Feminist Ethics: Theory and Practice*, Bloomington, IN: Indiana University Press.

Bud, R. (1993) *The Uses of Life: A History of Biotechnology*, Cambridge: Cambridge University Press.

Bunton, R. and Barrows, R. (1995) 'Consumption and health in the 'epidemiological' clinic of late modern medicine', in R. Bunton, S. Nettleton and R. Burrows (eds), *The Sociology of Health Promotion*, London: Routledge.

Bush, J. (2000) ' "It's just part of being a woman": cervical screening, the body and feminity', *Social Science and Medicine*, 50: 429–44.

Buskens, E., Grobbee, D.E., Hess, J. and Wladimiroff, J.W. (1995) 'Prenatal diagnosis of cogenital heart disease: prospects and problems', *European Journal of Obstetrics and Gynecology and Reproductive Biology* 60: 5–11.

Calafell, F.C. and Bertranpetit, J. (1993) 'The genetic history of the Iberian Peninsula: a simulation', *Current Anthropology* 34, 5: 735–45.

Calnan, M. (1987) *Health and Illness: The Lay Perspective*, London: Tavistock.

Casper, M. (1998) *The Making of the Unborn Patient: A Social Anatomy of Fetal Surgery*, New Brunswick, NJ: Rutgers University Press.

Chadwick, R. (2000) 'Ethical issues in psychiatric care: geneticisation and community care', *Act Psychiatrica Scandinavica* (Supplement 399) 101: 35–9.

Chandler, M. and Smith, A. (1998) 'Prenatal screening and women's perception of infant disability: A Sophie's Choice for every mother', *Nursing Inquiry* 5: 71–6.

Chappell, A.L. (1994) 'Disability, discrimination and the criminal justice system', *Critical Social Policy* 14, 3: 19–33.

Chard, T. (1996) 'Screening for Down's syndrome', *Clinical Endocrinology* 44, 15: 15.

Clarke, A. (1997a) 'Introduction', in A. Clarke and E. Parsons (eds), *Culture, Kinship and Genes: Towards Cross-Cultural Genetics*, Houndsmills: Macmillan.

—— (1997b) 'Prenatal genetic screening: paradigms and perspectives', in P.S. Harper and A.J. Clarke (eds), *Genetics, Society and Clinical Practice*, Oxford: BIOS Scientific Publishers Ltd.

Clarke, A.E. and Olsen, V.L. (1999) 'Revising, Diffracting, Acting', in A.E. Clarke and V.L. Olsen (eds), *Revisioning Women, Health and Healing: Feminist, Cultural and Technoscience Perspectives*, New York: Routledge.

Committee on Obstetric Practice (1994) American College of Obstetricians and Gynecologists Committee opinion 'Down syndrome screening', *International Journal of Gynecology and Obstetrics* 47: 186–90.

Concise Oxford Dictionary (1995) 9th edn, Oxford: Clarendon Press.

Cockburn, C. (1985) *Machinery of Dominance* London: Pluto Press.

Condit, C. and Condit, D. (2001) 'Blueprints and recipes: gendered metaphors for genetic medicine', *Journal of Medical Humanities* 22, 1: 39–55.

Conrad, P. (1997) 'Public eyes and private genes: historical frames, news constructions and social problems', *Social Problems* 44, 2: 139–54.

—— (1999) 'A Mirage of Genes', *Sociology of Health and Illness* 21, 2: 228–41.

Corea, G. (1985) *The Mother Machine*, London: The Women's Press.

Corea, G. *et al.* (1985) *Man-Made Women: How the New Reproductive Technologies affect Women*, London: Hutchinson.

Corker, M. and French, S. (1999) *Disability Discourse*, Buckingham: Open University Press.

Cowan, R.S. (1994) 'Women's roles in the history of amniocentesis and chorionic villi sampling', in K. Rothenberg, and E.J. Thomson (eds), *Women and Prenatal Testing: Facing the Challenges of Genetic Testing*, Columbus, OH: Ohio State University Press.

Cranor, C.F. (1994) 'Introduction', in C.F. Cranor (ed.), *Are Genes Us?: The Social Consequences of the New Genetics*, New Brunswick, NJ: Rutgers University Press.

Culliton, B.J. (1990) 'Mapping the terra incognita (Humani corporis)', *Science* 250: 210–12.

Cunningham, G.C. (2000) 'The genetics revolution – ethical, legal and insurance concerns', *Postgraduate Medicine* 108, 1: 193–204.

Cunningham, G.C. and Tompkinson, D.G. (1999) 'Cost and cost effectiveness of the triple marker prenatal screening program', *Genetics in Medicine* 1, 5: 199–206.

Currer, C. and Stacey, M. (1991) 'Introduction', in C. Currer and M. Stacey (eds), *Concepts of Health, Illness and Disease*, 2nd edn, New York and Oxford: Berg.

Davis, K. (1997) 'Embody-ing theory: beyond modernist and postmodernist readings of the body', in K. Davis (ed.), *Embodied Practices: Feminist Perspectives on the Body*, London: Sage.

Davison, C., Macintyre, S. and Davey Smith, G. (1994) 'The potential social impact of predictive genetic testing for susceptibility to common chronic diseases: a review and proposed research agenda', *Sociology of Health and Illness* 16, 3: 340–71.

Degener, T. (1990) 'Female self-determination between feminist claims and "voluntary" eugenics, between "rights" and 'ethics" ', *Issues in Reproductive and Genetic Engineering* 3, 2: 87–99.

De Gama, K. (1993) 'A brave new world? Rights, discourse and the politics of reproductive autonomy', *Journal of Law and Society* 20: 114–30.

Doyal, L. (1995) *What Makes Women Sick: Gender and the Political Economy of Health*, Houndsmill: Macmillan.

Dragonas, T. (2001) 'Whose fault is it? Shame and guilt for the genetic defect', in E. Ettorre (ed.), *Before Birth*, London: Ashgate.

Drake, R.F. (1996) 'A critique of the role of traditional charities', in L. Barton (ed.), *Disability and Society: Emerging Issues and Insights*, Harlow: Longman.

Duster, T. (1981) 'Intermediate steps between micro- and macro-integration: the case of screening for inherited disorders', in K. Knorr-Cetina and A.V. Cicourel (eds), *Advances in Social Theory and Methodology: Toward an Integration of Micro- and Macro-sociologies*, London: Routledge & Kegan Paul.

—— (1990) *Backdoor to Eugenics*, New York: Routledge.

Edwards, J., Franklin, S., Hirsch, E., Price, F. and Strathern, M. (1999) *Technologies of Procreation: Kinship in the Age of Assisted Reproduction*, 2nd edn, London: Routledge.

Ellison, P.T. (1990) 'Human ovarian function and reproductive ecology: new hypotheses', *American Anthropologist* 92, 4: 933–52.

Elshtain, J.B. (1991) 'Ethics in the women's movement', *The Annals of the American Academy of Political and Social Science* 515 (May): 126–50.

Ettorre, E. (1996) *The New Genetics Discourse in Finland: Exploring Experts' Views Within Surveillance Medicine*, Helsinki: Suomen Kuntaliitto (Association of Finnish Metropolitan Authorities) .

—— (ed.) (2001a) *Before Birth*, London: Ashgate.

—— (2001b) 'Experts' views on prenatal screening and diagnosis in Greece, the Netherlands, England and Finland', in E. Ettorre (ed.), *Before Birth*, London: Ashgate.

Exley, C. and Letherby, G. (2001) 'Managing a disrupted lifecourse: issues of identity and emotion work', *Health* 5, 1: 112–32.

Faden, R. (1994) 'Reproductive genetic testing, prevention and the ethics of mothering', in K. Rothenberg and E. J. Thomson (eds), *Women and Prenatal Testing: Facing the Challenges of Genetic Testing*, Columbus, OH: Ohio State University Press.

Farrant, W. (1985) 'Who's for amniocentesis?' in H. Homans (ed.), *The Sexual Politics of Reproduction*, Aldershot: Gower.

Faulkner, W. (2001) 'The technology question in feminism', *Women's Studies International Forum* 24, 1: 79–95.

Fears, R., Weatherall, D. and Post, G. (1999) 'The impact of genetics on medical education and training', *British Medical Bulletin* 55, 2: 460–70.

Featherstone, M. (1991) *Consumer Culture and Postmodernism*, London: Sage.

Finger, A. (1990) *Past Due: A Story of Disability, Pregnancy and Birth*, London: The Women's Press.

Foucault, M. (1973) *The Birth of the Clinic*, London: Tavistock Publications.

Fox Keller, E. (1995) *Refiguring Life: Metaphors of Twentieth Century Biology*, New York: Columbia University Press.

Frank, A. (1991) 'For a sociology of the body: an analytical review', in M. Featherstone, M. Hepworth and B. Turner (eds), *The Body: Social and Cultural Theory*, London: Sage.

—— (1995) *The Wounded Storyteller: Body, Illness and Ethics*, Chicago: University of Chicago Press.

Frankfort, E. (1972) *Vaginal Politics*, New York: Bantam Books.

Franklin, S. (1993) 'Postmodern procreation: representing reproductive practice', *Science as Culture* 13, 4: 522–61.

—— (1997) *Embodied Progress:A Cultural Account of Assisted Conception*, London: Routledge.

Freund, P.E.S. and McGuire, M.B. (1999) *Health, Illness and the Social Body: A Critical Sociology*, 3rd edn, Upper Saddle River, NJ: Prentice-Hall.

Friedman, M. (1992) 'Feminism and modern friendship: dislocating the community', in E. Browning-Cole and S. McQuinn-Coultrap (eds), *Explorations in Feminist Ethics: Theory and Practice*, Bloomington, IN: Indiana University Press.

Frye, M.M. (1983) *The Politics of Reality: Essays in Feminist Theory*, Freedom, CA: The Crossing Press.

Gatens, M. (1992) 'Power, bodies and difference', in M. Barrett and A. Phillips (eds), *Destabilizing Theory: Contemporary Feminist Debates*, Cambridge: Polity Press.

Genetic Interest Group (1995) *The Present Organisation of Genetic Services in the United Kingdom*, London: Genetic Interest Group.

Giddens, A. (1981) 'Agency, institution, and time-space analysis', in K. Knorr-Cetina and A.V. Cicourel (eds), *Advances in Social Theory and Methodology: Toward an Integration of Micro- and Macro-sociologies*, London: Routledge & Kegan Paul.

—— (1991) *The Consequences of Modernity*, Cambridge: Polity Press.

—— (1992) *The Transformation of Intimacy: Sexuality, Love and Eroticism in Modern Societies*, Cambridge: Polity Press.

—— (1994) 'Living in a Post Traditional Society', in U. Beck, A. Giddens and S. Lash (eds), *Reflexive Modernization: Politics, Tradition and Aesthetics in the Modern Social Order*, Stanford, CA: Stanford University Press.

Gilbert, N. and Mulkay, M. (1984) *Opening Pandora's Box: A Sociological Analysis of Scientist's Discourse*, Cambrige: Cambridge University Press.

Gilbert, S.F. (1997) 'Bodies of knowledge: biology and the intercultural university', in P. Taylor, S. Halfron and P. Edwards (eds), *Changing Life: Genomes, Ecologies, Bodies and Commodities*, Minneapolis: University of Minnesota Press.

Gillam, L. (1999) 'Prenatal diagnosis and discrimination against the disabled', *Journal of Medical Ethics* 25: 163–71.

Gilligan, C. (1982) *In a Different Voice: Psychological Theory and Women's Development*, Cambridge, MA: Harvard University Press.

Gilman, C. Perkins (1979) *Herland*, London: Women's Press.

Gimenez, M. (1991) 'The mode of reproduction in transition: a Marxist-feminist analysis of the effects of reproductive technologies', *Gender and Society* 5, 3: 334–50.

Gottweis, H. (1997) 'Genetic engineering, discourses of deficiency and the new politics of population', in P. Taylor, S. Halfron, and P. Edwards (eds), *Changing Life: Genomes, Ecologies, Bodies and Commodities*, Minneapolis: University of Minnesota Press.

Graham, H. and Oakley, A. (1991) 'Competing ideologies of reproduction: medical and maternal perspectives on pregnancy', in C. Currer and M. Stacey (eds), *Concepts of Health, Illness and Disease*, 2nd edn, New York and Oxford: Berg.

Green, J.M. (1990a) 'Prenatal screening and diagnosis: some psychological and social issues', *British Journal of Obstetrics and Gynaecology* 97: 1074–6.

—— (1990b) *Calming or Harming? A Review of Psychological Effects of Fetal Diagnosis on Pregnant Women*, London: The Galton Institute.

Gregg, R. (1995), *Pregnancy in a High-Tech Age: Paradoxes of Choice*, New York: New York University Press.

Habermas, J. (1984) *The Theory of Communicative Action*, Vol. 1, trans. T. McCarthy, Boston: Beacon Press.

Hales, G. (ed.) (1996) *Beyond Disability*, London: Sage.

Hallebone, E.L. (1992) 'Reproductive technology: repressive culture, and nongenetic mothers', *Reproductive and Genetic Engineering* 5, 3: 34–45.

Haraway, D. (1990) 'A manifesto for cyborgs: science, technology and socialist feminism in the 1980s', in L.J. Nicholson (ed.), *Feminism/Postmodernism*, London and New York: Routledge.

—— (1991) *Simians, Cyborgs and Women*, New York: Routledge.

—— (1997) 'The Virtual Speculum in the New World Order', *Feminist Review* 55: 22–72.

Harding, S. (1990) 'Feminism, science and the anti-enlightenment critiques', in L.J. Nicholson (ed.), *Feminism/Postmodernism*, London and New York: Routledge.

Hartouni, V. (1997) *Cultural Conceptions on Reproductive Technologies and the Remaking of Life*, Minneapolis: University of Minnesota Press.

Heckerling, P.S., Verp, M.S. and Hadro, T.A. (1994) 'Preferences of pregnant women for amniocentesis or chronic villus sampling for prenatal testing: comparison of patients' choices and those of a decision-analytic model', *Journal of Clinical Epidemiology* 47: 1215–28.

Henderson, B.J. and Maguire, B.T. (2000) 'Three lay models of disease inheritance', *Social Science and Medicine*, 50: 293–301.

Henn, W. (2000) 'Consumerism in prenatal diagnosis: a challenge for ethical guidelines', *Journal of Medical Ethics* 26: 444–6.

Heyd, D. (1992) *Genethics: Moral Issues in the Creation of People*, Berkeley, CA: University of California Press.

Higgs, P. (1997) 'The limits of medical knowledge', in G. Scambler (ed.), *Sociology as Applied to Medicine*, 4th edn, London: W.B. Saunders.

Hill Collins, P. (1999) 'Will the "real" mother please stand up? The logic of eugenics and American national family planning', in A.E. Clarke and V.L. Olsen (eds), *Revisioning Women, Health and Healing: Feminist, Cultural and Technoscience Perspectives*, New York and London: Routledge.

Holmes, H.B. (1992) 'A call to heal medicine', in H.B. Holmes and L. Purdy (eds), (1992) *Feminist Perspectives in Medical Ethics*, Bloomington, IN: Indiana University Press.

Holmes, H.B. and Purdy, L. (eds.) (1992) *Feminist Perspectives in Medical Ethics*, Bloomington, IN: Indiana University Press.

Howson, A. (1998) 'Embodied obligation: the female body and health surveillance', in S. Nettleton and J. Watson (eds), *The Body in Everyday Life*, London: Routledge.

Hubbard, R. (1985) 'Genomania and health', *American Scientist* 83: 8–10.

—— (1986) 'Eugenics and prenatal testing', *International Journal of Health Services* 16, 2: 227–42.

Hubbard, R. and E. Wald (1993) *Exploding the Gene Myth*, Boston: Beacon Press.

Hughes, G. (1998) 'A suitable case for treatment? Constructions of disability', in E. Saraga (ed.), *Embodying the Social: Constructions of Difference*, London: Routledge.

Hughes, A. and Witz, A. (1997) 'Feminism and the matter of bodies: from de Beauvoir to Butler', *Body and Society* 3, 1: 47–60.

Huyssen, A. (1990) 'Mapping the postmodern', in L.J. Nicholson (ed.), *Feminism/Postmodernism*, London and New York: Routledge.

I Ching, or Book of Changes (1951) trans. Richard Wilhelm, London: Routledge.

Jewkes, R.K and Wood, K. (1999) 'Problematizing pollution: dirty wombs, ritual pollution and pathological processes, *Medical Anthropology* 18, 163–86.

Jones, J. (2000) 'The Tuskegee syphilis experiment', in P. Brown (ed.), *Perspectives in Medical Sociology*, 3rd edn, Prospect Heights, IL: Waveland Press.

Jonsen, A.R. (1996) 'The impact of mapping the human genome on the patient–physician relationship', in T. Murray, M.A. Rothstein and R.F. Murray (eds), *The Human Genome Project and the Future of Health Care*, Bloomington, IN: Indiana University Press.

Kabakian-Khasholian, T., Campbell, O., Shediac-Rizkallah, M. and Ghorayeb, F. (2000) 'Women's experience of maternity care: satisfaction or passivity?', *Social Science and Medicine* 51: 103–13.

Katz Rothman, B.K. (1994) *The Tentative Pregnancy: Amniocentesis and the Sexual Politics of Motherhood*, London: Pandora.

—— (1995) 'Of maps and imaginations: sociology confronts the genome', *Social Problems* 42, 1: 1–10.

—— (1996) 'Medical sociology confronts the human genome', *Medical Sociology News*, 22 December, 1: 23–35.

—— (1998a) *Genetic Maps and Human Imaginations*, New York: W.W. Norton.

—— (1998b) 'A sociological sceptic in the brave new world', *Gender and Society* 12: 501–4.

Keith, L. and Morris, J. (1995) 'Easy targets: a disability rights perspective on the "children as carers"', *Critical Social Policy* 15, 2/3: 36–57.

Kelly, M.P. and Field, D. (1996) 'Medical sociology, chronic illness and the body', *Sociology of Health and Illness* 18, 2: 241–57.

Kerr, A. and Cunningham-Burley, S. (2000) 'On ambivalence and risk: reflexive modernity and the new human genetics', *Sociology* 34, 2: 283–304.

Kerr, A., Cunningham-Burley, S. and Amos, A. (1997) 'The new genetics: professionals' discursive boundaries', *Sociological Review* 45, 2: 297–303.

—— (1998) 'Eugenics and the new genetics in Britain: examining contemporary professionals' accounts', *Science Technology and Human Values* 23: 175–98.

Kimbrell, A. (1993) *The Human Body Shop*, San Francisco: Harper.

Klassen, P.E. (2001) 'Sacred maternities and postmedical bodies: religion and nature in contemporary home birth', *Signs: Journal of Women in Culture and Society* 26, 3: 775–810.

Kleinman, A. (1991) 'Concepts and a model for the comparison of medical systems as cultural systems', in C. Currer and M. Stacey (eds), *Concepts of Health, Illness and Disease*, 2nd edn, New York and Oxford: Berg.

Koch, L. and Stermerding, D. (1994) 'The sociology of entrenchment: a cystic fibrosis test for everyone?', *Social Science and Medicine* 39, 9: 1211–20.

Kuhse, H. and Singer, P. (1999) 'Introduction', in H. Kuhse and P. Singer (eds), *Bioethics: An Anthology*, Oxford: Blackwell Publishers.

Kuller, J.A and Laifer, S.A. (1995) 'Contemporary approaches to prenatal diagnosis', *American Family Physician* 52, 8: 2277–83.

Kuppermman, M., Nease, R.F., Learman, L.A., Gates, E., Blumberg, B. and Washington, A.E. (2000) 'Procedure related miscarriages and Down syndrome-effected births implications for prenatal testing based on women's preferences', *Obstetrics and Gynecology* 96, 4: 511–16.

Landsman, G.H. (1998) 'Reconstructing motherhood in the age of "perfect" babies: mothers of infants and toddlers with disabilities', *Signs: Journal of Women in Culture and Society* 24, 1: 9–98.

Latour, B. and Woolgar, S. (1986) *Laboratory Life: the Construction of Scientific Facts*, 2nd edn, Princeton, NJ: Princeton University Press.

Laurén, M., Petrogiannis, K, Valassi-Adam, E. and Tymstra, T. (2001) 'Prenatal diagnosis in the lay press and professional journals in Finland, Greece and the Netherlands', in E. Ettorre (ed.), *Before Birth*, London: Ashgate.

Lebacqz, K. (1991) 'Feminism and bioethics: an overview', *Second Opinion*, 17: 11–25.

Leder, D. (1992) 'A tale of two bodies: the Cartesian corpse and the lived body', in D. Leder (ed.), *The Body in Medical Thought and Practice*, Dordrecht: Kluwer Academic Publishers.

Lee, R. and Morgan, D. (1989) 'A lesser sacrifice? Sterilization and Mentally Handicapped', in R. Lee and D. Morgan (eds), *Birthrights, Law and Ethics at the Beginnings of Life*, New York and London: Routledge.

Lewin, E. (1985) 'By design: reproductive strategies and the meaning of moth-erhood', in H. Homans (ed.), *The Sexual Politics of Reproduction*, Aldershot: Gower.

Lilford, R.J. and Thornton, J.G. (1996) 'Prenatal screening vouchers', *Journal of the Royal Society of Medicine* 89: 130–1.

Lilford, R.J., Vanderkerckhove, P. and Thornton, J.G. (1994) 'Decision analy-sis in clinical genetics', *Bailliere's Clinical Obstetrics and Gynaecology* 8, 3: 625–42.

Lippman, A. (1992) 'Led astray by genetic maps: the cartography of the human genome and health care', *Social Science and Medicine*, 35: 1469–76.

—— (1994) 'The genetic construction of prenatal testing', in K. Rothenberg and E. J. Thomson (eds), *Women and Prenatal Testing: Facing the Chal-lenges of Genetic Testing*, Columbus, OH: Ohio State University Press.

—— (1999) 'Choice as a risk to women's health', *Health, Risk and Society* 1, 3: 281–92.

Little, Margaret Olivia (1996) 'Why a feminist approach to bioethics?', *Kennedy Institute of Ethics Journal* 6, 1: 1–18.

Locke, S. (2001) 'Sociology and the public understanding of science: from rationalization to rhetoric', *British Journal of Sociology* 52, 1: 1–18.

Lomas, J. (1998) 'Social capital and health: implications for public health and epidemiology', *Social Science and Medicine* 47: 1181–8.

Lorber, J. (1994) *Paradoxes of Gender*, New Haven, CT: Yale University Press.

—— (1997) *Gender and the Social Construction of Illness*, Thousand Oaks, CA and London: Sage Publications.

Lorber, J. and Yancey Martin, P. (1997) 'The socially constructed body: insights from feminist theory', in P. Kivisto (ed.), *Illuminating Social Life*, Thousand Oaks, CA: Pine Forge Press.

Loukopoulos, D. *et al.* (1990) 'Prenatal diagnosis of thalassemia and of sickle cell syndromes in Greece', *Annuals of the New York Academy of Science* 612: 226–36.

Lundin, S. (1997) 'Visions of the body', in S. Lundin and M. Ideland (eds), *Gene Technology and the Public*, Lund: Nordic Academic Press.

Lupton, (1994) *Medicine as Culture: Illness, Disease and the Body in Western Societies*, London: Sage.

Mahowald, M.B. (1994) 'Reproductive genetics and gender justice', in K. Rothenberg and E.J. Thomson (eds), *Women and Prenatal Testing: Facing the Challenges of Genetic Testing*, Columbus, OH: Ohio State University Press.

Malone, P.S.J. (1996) 'Antenatal diagnosis of renal tract anomalies: has it increased the sum of human happiness?', *Journal of the Royal Society of Medicine* 89, 3: 155–8.

Markens, S., Browner, C.H. and Press, N. (1999) 'Because of the risks: how US women account for refusing prenatal screening', *Social Science and Medicine* 49, 3: 359–69.

Marshall, H. and Woollett, A. (2000) 'Fit to reproduce? The regulative role of pregnancy texts', *Feminism and Psychology* 10, 3: 351–66.

Marteau, T.M. (1989) 'Psychological costs of screening may sometimes be bad enough to undermine the benefits of screening', *British Medical Journal* 299 (29 August): 527.

—— (1995) 'Towards informed decisions about prenatal testing: a review', *Prenatal Diagnosis* 15: 1215–26.

Marteau, T.M. and Drake, H. (1995) 'Attributions of disability: the influence of genetic screening', *Social Science and Medicine* 40, 8: 1127–32.

Marteau, T.M., Mitchie, S., Drake, H. and Bobrow, M. (1995) 'Public attitudes towards the selection of desirable characteristics in children', *Journal of Medical Genetics* 32: 796–98.

Marteau, T., Shaw, R. and Slack, J. (1994), *Routine Prenatal Testing for Fetal Abnormalities*, London: Medical Research Council.

Martin, B. (1992) 'Elements of a Derridean social theory' in A. Dallery, C.E. Scott and H. Roberts (eds), *Ethics and Danger: Essays on Heidegger and Continental Thought*, Albany, NY: State University of New York Press.

Martin, E. (1992) *The Woman in the Body: A Cultural Analysis of Reproduction*, Boston: Beacon Press.

McCaughey, M. (1993) 'Evolution, ethics and the search for uncertainty', *Science As Culture* 4, 2: 212–43.

McDowell, L. (1999) *Gender, Identity and Place: Using Feminist Geographies*, Cambridge: Polity Press.

McNeil, M. (1993) 'New reproductive technologies: dreams and broken promises', *Science as Culture* 3, 4, 17: 483–505.

Minden, S. (1987) 'Patriarchal designs: the genetic engineering of human embryos', in P. Spallone and D.L. Steinberg (eds), *Made to Order: The Myth of Reproductive and Genetic Engineering*, Oxford and New York: Pergamon Press.

Ministry of Welfare, Health and Sport (1996) *The Population Screening Act*, Rijswijk.

Morgall, J. M. (1993) *Technology Assessment: A Feminist Perspective*, Philadelphia: Temple University Press.

Morris, J. (1991) *Pride Against Prejudice: Transforming Attitudes to Disability*, London: The Womens' Press.

Murray, T.H. and Livny, E. (1995) 'The human genome project: ethical and social implications', *Bulletin of the Medical Library Association* 83, 1: 14–21.

Nelkin, D. and Lindee, M.S. (1995a) *The DNA Mystiques: The Gene as a Cultural Icon*, New York: W.H. Freeman and Company.

—— (1995b) 'The media-ted gene: stories of gender and race', in J. Terry and J. Urla (eds), *Deviant Bodies: Critical Perspectives on Difference in Science and Popular Cultures*, Bloomington, IN: Indiana University Press.

Nettleton, S. and J. Watson (eds) (1998) *The Body in Everyday Life*, London: Routledge.

Newberger, D.S. (2000) 'Down syndrome: prenatal risk assessment and diagnosis', *American Family Physician* 62, 4: 825–32.

Newman, K. (1996) *Fetal Positions: Individualism, Science and Visuality*, Stanford, CA: Stanford University Press.

Nicholas, B. (2001) 'Exploring a moral landscape: genetic science and ethics', *Hypatia* 16, 1: 45–63.

Nielsen, T.H. (1997) 'Modern biotechnology – sustainability and integrity', in S. Lundin and M. Ideland (eds), *Gene Technology and the Public: An Interdisciplinary Perspective*, Lund: Nordic Academic Press.

Novas, C. and Rose, N. (2000) 'Genetic risk and the birth of the somatic individual', *Economy and Society* 29, 4: 485–513.

Oakley, A. (1984) *The Captured Womb: A History of the Medical Care of Pregnant Woman*, Oxford: Basil Blackwell.

O'Connell, M.P., Holding, S., Morgan, R.J. and Lindow, S.W. (2000) 'Biochemical screening for Down's syndrome: patients' perception of risk', *International Journal of Gynecology and Obstetrics* 68, 3: 215–18.

Oliver, M. (1996) *Understanding Disability: From Theory to Practice*, London: Macmillan.

Oliver, M. and Barnes, C. (1998) *Disabled People and Social Policy: From Exclusion to Inclusion*, London and New York: Longman.

Overall, C. (1987) *Ethics and Human Reproduction*, Boston: Allen and Unwin.

Overboe, J. (1999) 'Difference in itself: validating disabled people's lived experiences', *Body and Society* 5, 4: 17–29.

Parker, M. and Hope, T. (2000) 'Medical ethics in the 21st century', *Journal of Internal Medicine* 248, 1: 1–6.

Parsons, E. and Atkinson, P. (1993) 'Genetic risk and reproduction', *Sociological Review* 41: 679–706.

Paul, D.B. (1992) 'Eugenic anxieties, social realities and political choices', *Social Research* 59, 3: 663–83.

Petersen, A. (1998) 'The new genetics and the politics of public health', *Critical Public Health* 8: 59–71.

—— (2001) 'Biofantasies: genetics and medicine in the print new media', *Social Science and Medicine* 52: 1255–68.

Petrogiannis, K., Tymstra,T., Jallinoja, P. and Ettorre, E. (2001) 'Review of policy, law and ethics', in E. Ettorre (ed.), *Before Birth*, London: Ashgate.

Pinder, R. (1995) 'Bringing back the body without the blame? The experience of ill and disabled people at work', *Sociology of Health and Illness* 17, 5: 605–31.

Press N. and Browner, C.H. (1997) 'Why women say yes to prenatal diagnosis', *Social Science and Medicine* 45, 7: 979–89.

Price, F. (1999) 'Beyond expectation: child practices and clinical concerns', in J. Edwards, S. Franklin, E. Hirsch, F. Price and M. Strathern, *Technologies of Procreation: Kinship in the Age of Assisted Reproduction*, 2nd edn, London: Routledge.

Priestley, M. (1995) 'Dropping "E"s: the missing link in quality assurance for disabled people', *Critical Social Policy* 15, 2/3: 7–21.

Purdy, L. (1992) 'A call to heal ethics', in H.B. Holmes, and L. Purdy (eds), *Feminist Perspectives in Medical Ethics*, Bloomington, IN: Indiana University Press.

—— (1996) *Reproducing Persons: Issues in Feminist Bioethics*, Ithaca, NY: Cornell University Press.

—— (1999) 'Genetics and reproductive risk: can having children be immoral?', in H. Kuhse and P. Singer (eds), *Bioethics: An Anthology*, Oxford: Blackwell.

Qureshi, N. and Raeburn, J.A. (1993) 'Clinical genetics meets primary care', *British Medical Journal* 307: 816–7.

Rabinow, P. (1999) *French DNA: Trouble in Purgatory*, Chicago: University of Chicago Press.

Rapp, R (1994) 'Women's response to prenatal diagnosis: a sociocultural perspective', in K. Rothenberg and E.J. Thomson (eds), *Women and Prenatal Testing: Facing the Challenges of Genetic Testing*, Columbus, OH: Ohio State University Press.

—— (1998) 'Refusing prenatal diagnosis: the uneven meaning of biosciences in a multicultural world', in R. Davis-Floyd and J. Dumit (eds), *Cyborg Babies: From Techno-Sex to Techno-Tots*, New York and London: Routledge.

—— (1999) *Testing Women, Testing the Fetus: The Social Impact of Amniocentesis*, New York and London: Routledge.

Rausch, D.N., Lambert-Messerlian, G.M. and Canick, J.A. (2000) 'Participation in maternal serum screening following positive results in a previous pregnancy', *Journal of Medical Screening* 7, 1: 4–6.

Reed, E. (1978) *Sexism and Science*, New York: Pathfinder Press.

Reid, M. (1991) *The Diffusion of Four Prenatal Screening Tests Across Europe*, London: King's Fund Centre for Health Services Development.

Resta, R. (1999) 'A brief history of the pedigree in human genetics', in R.A. Peel (ed.), *Human Pedigree Studies*, London: The Galton Institute.

Rhodes, R. (1998) 'Genetic links, family ties, and social bonds: rights and responsibilities in the face of genetic knowledge', *Journal of Medicine and Philosophy* 23, 1: 10–30.

Rich, A. (1977) 'The theft of childbirth', in Claudia Dreifus (ed.), *Seizing Our Bodies: The Politics of Women's Health*, New York: Vintage Books.

Richards, M. (1997) 'It runs in the family: lay knowledge about inheritance', in A. Clarke and E. Parsons (eds), *Culture, Kinship and Genes: Towards Cross-Cultural Genetics*, Houndsmills: Macmillan.

Richards, M.P.M. (1993) 'The new genetics: some issues for social scientists', *Sociology of Health and Illness* 15, 5: 567–86.

Rose, S. and Rose, H. (1969) *Science and Society*, Harmondsworth: Penguin Books.

Rosner, M. and Johnson, T.R. (1995) 'Telling stories: metaphors of the human genome project', *Hypatia* 10, 4: 104–29.

Roth, J. (1963) *Timetables: Structuring the Passage of Time in Hospital Treatment and Other Careers*, New York: The Bobbs Merrill Company Inc.

Rothenberg, K. (1996) 'Feminism, law and ethics', *Kennedy Institute of Ethics Journal* 6, 1: 69–84.

Rothenberg, K. and Thomson, E.J. (1994) 'Women and prenatal testing', in K. Rothenberg and E.J. Thomson (eds), *Women and Prenatal Testing: Facing the Challenges of Genetic Testing*, Columbus, OH: Ohio State University Press.

Rudinow Saetnan, A. (1996) 'Contested meanings of gender and technology in the Norwegian ultrasound screening debate', *The European Journal of Women's Studies* 3, 1: 55–75.

Russ, J. (1985) *The Female Man*, London: The Women's Press.

Santalahti, P. (1998) *Prenatal Screening in Finland*, Helsinki: National Research and Development Centre for Welfare and Health.

Sbisa, M. (1996) 'The feminine subject in discourse about childbirth', *The European Journal of Women's Studies* 3, 4: 363–76.

SGOMSEC (Scientific Group on Methodologies for the Safety Evaluation of Chemicals) (1996) *Proceedings of the 12th SGOMSEC Workshop*, Rutgers, NJ: SGOMSEC.

Shakespeare, T. (1995) 'Back to the future? New genetics and disabled people', *Critical Social Policy* 15: 22–35.

Sharp, L.A. (2000) 'The commodification of the body and its parts', *Annual Review of Anthropology*, 29: 287–328.

Shildrick, M. (1997) *Leaky Bodies and Boundaries: Feminism, Postmodernism and (Bio)ethics*, London and New York: Routledge.

Shilling, C. (1993) *The Body and Social Theory*, London: Sage Publications.

—— (1998) 'The undersocialised conception of the (embodied) agent in modern sociology', *Sociology* 31: 737–54.

Shore, C. (1992) 'Virgin births and sterile debates', *Current Anthropology* 33, 3: 295–314.

Skupski, D.W., Chervenak, F.A., McCullough, L.B. (1994) 'Is routine ultrasound screening for all patients?', *Clinics in Perinatology* 21, 4: 707–22.

Spallone, P. (1989) *Beyond Conception: The New Politics of Reproduction*, London: Macmillan.

—— (1995) 'Bad conscience and collective unconscious: science, discourse and reproductive technology', in B. Rosenback and R.M. Schott (eds), *For Plantning, kon og teknologi*, Copenhagen: Museum Tusculanums Forlag.

Spallone, P. and Steinberg, D.L. (1987a) 'Introduction', in P. Spallone and D.L. Steinberg (eds), *Made to Order: The Myth of Reproductive and Genetic Engineering*, Oxford and New York: Pergamon Press.

—— (1987b) *Made to Order: The Myth of Reproductive and Genetic Engineering*, Oxford and New York: Pergamon Press.

Spallone, P., Wilkie, T., Ettorre, E., Haimes, E., Shakespeare, T. and Stacey, M. (2000) 'Putting sociology on the bioethics map', in J. Eldridge, J. MacInnes, S. Scott, C. Warhurst and A. Witz (eds), *For Sociology*, Durham: Sociology Press/British Sociological Association.

Spencer, K. (1999a) 'Second trimester prenatal screening for Down's syndrome using alpha-fetoprotein and free beta hCG: a seven year review', *British Journal of Obstetrics and Gynaecology* 106, 12: 1287–94.

—— (1999b) 'Accuracy of Down's syndrome risks produced in a prenatal screening program', *Annals of Clinical Biochemistry* 36, 1: 101–3.

Spencer, K., Spencer, C.E., Power, M., Moakes, A. and Nicolaides, K.H. (2000) 'One stop clinic for assessment of risk for fetal anomalies: a report of the first year of prospective screening for chromosomal anomalies in the first trimester', *British Journal of Obstetrics and Gynaecology* 107, 10: 1271–5.

Spurr, N., Darvasi, A., Terrentt, J. and Jazwinska, L. (1999) 'New technologies and DNA resourcres for high throughput biology', *British Medical Bulletin* 55, 2: 309–24.

Stabile, C.A. (1994) *Feminism and the Technological Fix*, Manchester: Manchester University Press.

Stacey, M. (1988) *The Sociology of Health and Healing*, London: Unwin Hyman Ltd.

—— (1992) *Changing Human Reproduction: Social Science Perspectives*, London: Sage.

Stanworth, M. (ed.) (1987) *Reproductive Technologies: Gender, Motherhood and Medicine*, Cambridge: Polity Press.

Steinberg, D.L. (1992) 'Genes and racial hygiene: studies of science under National Socialism', *Science as Culture* 3,4 (Part 1): 116–29.

—— (1997) *Bodies in Glass: Genetics, Eugenics, Embryo Ethics*, Manchester and New York: Manchester University Press.

Stormer, N. (2000) 'Prenatal space', *Signs: Journal of Women in Culture and Society* 26, 1: 109–44.

Strathern, M. (1997) 'The work of culture: an anthrological perspective', in A. Clarke and E. Parsons (eds), *Culture, Kinship and Genes*, Houndsmill: Macmillan.

—— (1999) 'Postscript: a relational view', in J. Edwards, S. Franklin, E. Hirsch, F. Price and M. Strathern, *Technologies of Procreation: Kinship in the Age of Assisted Reproduction*, 2nd edn, London: Routledge.

Stronach, I. and Allan, J. (1999) 'Joking with disability: what's the difference between the comic and the tragic in disability discourses', *Body and Society* 5, 4: 31–45.

Tarlov, A.R. (1996) 'Social determinants of health: the sociobiological translation', in D. Blane, E. Brunner and R. Wilkinson (eds), *Health and social organization: towards a health policy for the twenty-first century*, London and New York: Routledge.

Taylor, J.S. (1993) 'The public foetus and the family car', *Science as Culture* 3, 17: 601–18.

—— (2000) 'Of sonograms and baby prams: prenatal diagnosis, pregnancy and consumption', *Feminist Studies* 26, 2: 391–418.

Terry, J. and Urla, J. (eds) (1995) *Deviant Bodies: Critical Perspectives On Difference in Science and Popular Cultures*, Bloomington, IN: Indiana University Press.

Thomas, C. (1997) 'The baby and the bath water: disabled women and motherhood in a social context', *Sociology of Health and Illness* 19, 5: 622–43.

Tietjens Meyers, D. (2001) 'The rush to motherhood – pronatalist discourse and women's autonomy', *Signs: Journal of Women in Culture and Society* 26, 3: 735–72.

Tomm, W. (1992) 'Ethics and self-knowing: the satisfaction of desire', in E. Browning-Cole and S. McQuinn-Coultrap (eds), *Explorations in Feminist Ethics: Theory and Practice*, Bloomington, IN: Indiana University Press.

Turner, B. (1987) *Medical Power and Social Knowledge*, London: Sage.
—— (1992) *Regulating Bodies: Essays in Medical Sociology*, London and New York: Routledge.
—— (1996) *The Body and Society*, 2nd edn, London: Sage.
Tymstra, T. (1991) 'Prenatal diagnosis, prenatal screening and the rise of the tentative pregnancy', *International Journal of Technology Assessment in Health Care* 7, 4: 509–16.
Urla, J. and Terry, J. (1995) 'Introduction: mapping embodied deviance', in J. Terry and J. Urla (eds), *Deviant Bodies: Critical Perspectives on Difference in Science and Popular Cultures*, Bloomington, IN: Indiana University Press.
Urla, J. and Swedlund, A. (1995) 'The anthropomentry of barbie: unsettling ideals of the feminine body in popular culture', in J. Terry and J. Urla (eds), *Deviant Bodies: Critical Perspectives on Difference in Science and Popular Cultures*, Bloomington, IN: Indiana University Press.
Van Dijck, T. (1998) *Imagenation: Popular Images of Genetics*, New York: New York University Press.
Verlinsky, Y. *et al.* (1997) 'Prepregnancy genetic testing for age-related aneuploidies by polar body analysis', *Genetic Testing* 1, 4: 231–5.
Wajcman, J. (1991) *Feminism Confronts Technology*, University Park, PA: Pennsylvania State University Press.
Wald, N.J., Densem, J.W., George, L., Muttukrishna, S., and Knight, P.G. (1996) 'Prenatal screening for Downs's syndrome using Inhibin-A as a serum marker', *Prenatal Diagnosis* 16, 143–153.
Ward, C.M. (1994) 'An ethical and legal perspective on foetal surgery', *British Journal of Plastic Surgery* 47, 6: 411–18.
Warren, V.L. (1992) 'Feminist directions in medical ethics', in H.B. Holmes and L. Purdy (eds), *Feminist Perspectives in Medical Ethics*, Bloomington, IN: Indiana University Press.
Wendell, S. (1992) 'Toward a feminist theory of disability', in H.B. Holmes and L. Purdy (eds), *Feminist Perspectives in Medical Ethics*, Bloomington, IN: Indiana University Press.
—— (1996) *The Rejected Body: Feminist Philosophical Reflections on Disability*, New York: Routledge.
Wertz, D.C. and Fletcher, J.C. (1993) 'Prenatal diagnosis and sex selection in 19 nations', *Social Science and Medicine* 37, 11: 1359–66.
—— (1998) 'Ethical and social issues in prenatal sex selection: a survey of geneticists in 37 nations', *Social Science and Medicine* 46, 2: 255–73.
Wertz, D.C. and Gregg, R. (2000) 'Genetic services in a social, ethical and policy context: a collaboration between consumers and providers', *Journal of Medical Ethics*, 26: 261–5.
Wetherall, D.J. (1991) *The New Genetics and Clinical Practice*, Oxford: Oxford University Press.
Wilkinson, R.G. (1996) *Unhealthy Societies: The Afflictions of Inequality*, London and New York: Routledge.
Williams, S. (1998) 'Capitalising on emotions? Rethinking the inequalities in health debate', *Sociology* 32: 121–39.

Williams, S. and Bendelow, G. (1998) *The Lived Body: Sociological Themes, Embodied Issues*, London: Routledge.

Witz, A. (2000) 'Whose body matters? Feminist sociology and the corporeal turn in sociology and feminism', *Body and Society*, 6, 2: 1–24.

Wright, S. (1994) *Molecular Politics: Developing American and British Regulatory Policy for Genetic Engineering, 1972–82*, Chicago and London: University of Chicago Press.

Zerubavel, E. (1982) 'The standardization of time: a sociohistorical perspective', *American Journal of Sociology* 88, 1: 1–23.

—— (1997) *Social Mindscapes: An Invitation to Cognitive Sociology*, Cambridge, MA: Harvard University Press.

Index